Alpheus. S. Packard

**Our Common Insects**

A popular account of the insects of our fields, forests, gardens and houses

Alpheus. S. Packard

**Our Common Insects**
*A popular account of the insects of our fields, forests, gardens and houses*

ISBN/EAN: 9783337081843

Printed in Europe, USA, Canada, Australia, Japan

Cover: Foto ©berggeist007 / pixelio.de

More available books at **www.hansebooks.com**

# OUR COMMON INSECTS.

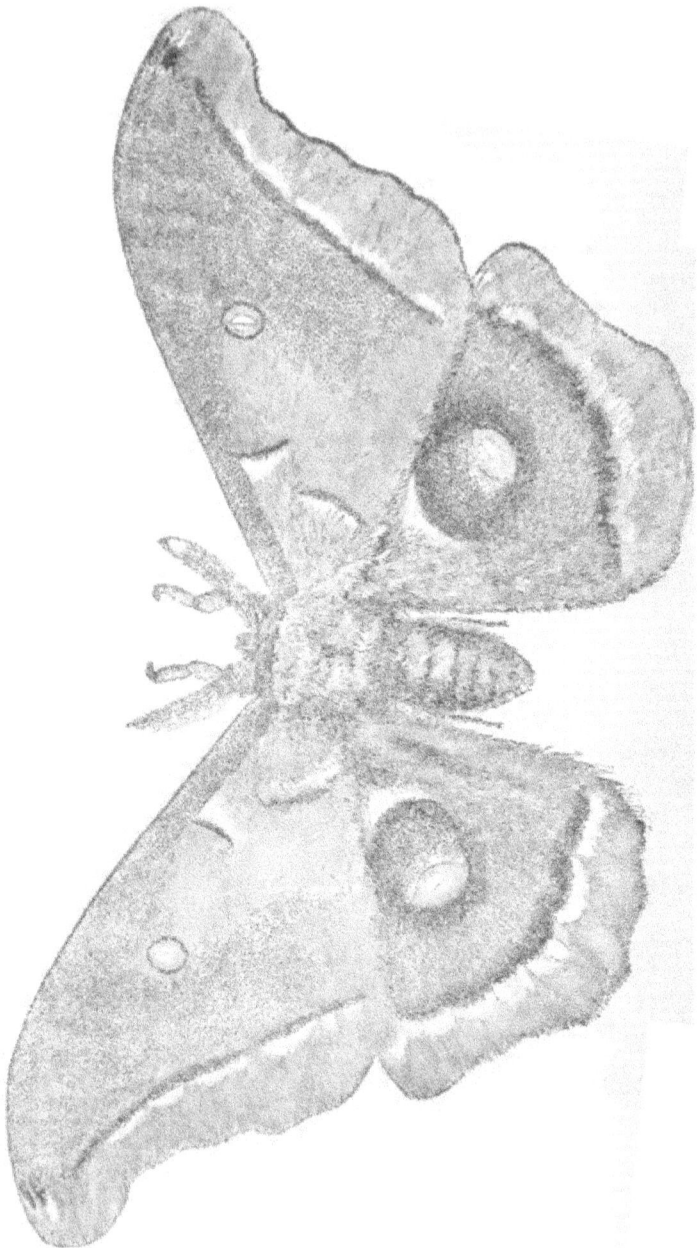

OUR

# COMMON INSECTS.

## A POPULAR ACCOUNT OF THE INSECTS

### OF OUR

## Fields, Forests, Gardens and Houses.

Illustrated with 4 Plates and 268 Woodcuts.

By

## A. S. PACKARD, Jr.,

Author of "A GUIDE TO THE STUDY OF INSECTS."

SALEM.

NATURALISTS' AGENCY.

BOSTON: Estes & Lauriat.   NEW YORK: Dodd & Mead.

1873.

# PREFACE.

This little volume mainly consists of a reprint of a series of essays which appeared in the "American Naturalist" (Vols. i–v, 1867–71). It is hoped that their perusal may lead to a better acquaintance with the habits and forms of our more common insects. The introduction was written expressly for this book, as well as Chapter XIII, "Hints on the Ancestry of Insects." The scientific reader may be drawn with greater interest to this chapter than to any other portion of the book. In this discussion of a perhaps abstruse and difficult theme, his indulgence is sought for whatever imperfections or deficiencies may appear. Our systems of classification may at least be tested by the application of the theory of evolution. The natural system, if we mistake not, is the genealogy of organized forms; when we can trace the latter, we establish the former. Considering how much naturalists differ in their views as to what is a natural classification, it is not strange that a genealogy of animals or plants seems absurd to many. To another generation of naturalists it must, perhaps, be left to decide whether to attempt the one is more unphilosophical than to attempt the other.

Most of the cuts have already appeared in the "Guide to the Study of Insects" and the "American Naturalist," where their original sources are given, while a few have been kindly contributed by Prof. A. E. Verrill, the Boston Society of Natural History, and Prof. C. V. Riley, and three are original.

SALEM, June, 1873.

# OUR COMMON INSECTS.

## INTRODUCTORY.

*What is an Insect?* When we remember that the insects alone comprise four-fifths of the animal kingdom, and that there are upwards of 200,000 living species, it would seem a hopeless task to define what an insect is. But a common plan pervades the structure of them all. The bodies of all insects consist of a succession of rings, or segments, more or less hardened by the deposition of a chemical substance called chitine; these rings are arranged in three groups: the head, the thorax, or middle body, and the abdomen or hind body. In the six-footed insects, such as the bee, moth, beetle or dragon fly, four of these rings unite early in embryonic life to form the head; the thorax consists of three, as may be readily seen on slight examination, and the abdomen is composed either of ten or eleven rings. The body, then, seems divided or *insected* into three regions, whence the name *insect*.

The head is furnished with a pair of antennæ, a pair of jaws (mandibles), and two pairs of maxillæ, the second and basal pair being united at their base to form the so-called labium, or under lip. These four pairs of appendages represent the four rings of the head, to which they are appended in the order stated above.

A pair of legs is appended to each of the three rings of the thorax; while the first and second rings each usually carry a pair of wings.

The abdomen contains the ovipositor; sometimes, as in the bees and wasps, forming a sting. In the spiders (Fig. 1), however, there are no antennæ, and the second maxillæ, or labium, is wanting. Moreover, there are four pairs of legs. The centipedes (Fig. 2, a Myriopod) also differ from the rest of the insects in having an indefinite number of abdominal rings, each bearing a pair of legs.

On examining the arrangement of the parts

1. Spider (Tegenaria).

within, we find the nervous cord, consisting of two chains of swellings, or nerve-knots, resting upon the floor or under side of the body; and the heart, or dorsal vessel, situated just under the skin of the back; and in looking at living caterpillars, such as the cut-worm, and many thin-skinned aquatic larvæ, we can see this long tubular heart pulsating about as often as our own heart, and when the insect is held against its will, or is agitated, the rapidity of the pulsations increases just as with us.

Insects do not breathe as in the higher animals by taking the air into the mouth and filling the lungs, but there are a series of holes or pores along the side of the body, as seen in the grub of the humble bee, through which the air enters and is conveyed to every part of the body by an immense number of air tubes. (Fig. 3, air tubes, or tracheæ, in the cau-

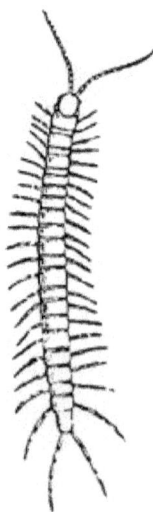

2. Centipede.

dal appendage of the larva of a dragon fly). These air tubes
are everywhere bathed by the blood, by which the latter becomes
oxygenated.

Indeed the structure of an insect is entirely different from
that of man or the
quadrupeds, or any
other vertebrate ani-
mal, and what we
call head, thorax,

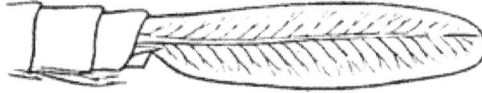

3. Caudal appendage of larva of Agrion.

abdomen, gills, stomach, skin, or lungs, or jaws, are called so
simply for convenience, and not that they are made in the same
way as those parts in the higher animals.

An insect differs from a horse, for example, as much as a
modern printing press differs from the press Franklin used.
Both machines are made of iron, steel, wood, etc., and both
print; but the plan of their structure differs throughout, and
some parts are wanting in the simpler press which are present
and absolutely essential in the other. So with the two sorts
of animals; they are built up originally out of protoplasm, or
the original jelly-like germinal matter, which fills the cells composing their tissues, and nearly the same chemical elements
occur in both, but the mode in which these are combined, the
arrangement of their products: the muscular, nervous and skin
tissues, differ in the two animals. The plan of structure, namely,
the form and arrangement of the body walls, the situation of
the appendages to the body, and of the anatomical systems within,
i. e., the nervous, digestive, circulatory, and respiratory systems,
differ in their position in relation to the walls of the body.
Thus while the two sorts of animals reproduce their kind, eat,
drink and sleep, see, hear and smell, they perform these acts by
different kinds of organs, situated sometimes on the most opposite parts of the body, so that there is no comparison save in
the results which they accomplish; they only agree in being
animals, and in having a common animal nature.

*How Insects Eat.* The jaws of insects (Fig. 4) are horny processes situated on each side of the mouth. They are variously

toothed, so as to
tear the food,
and move hori-
zontally instead

4. Different forms of jaws.

of up and down

as in the horse. The act of taking the food, especially if the
insect be carnivorous in its habits, is quite complex, as not only

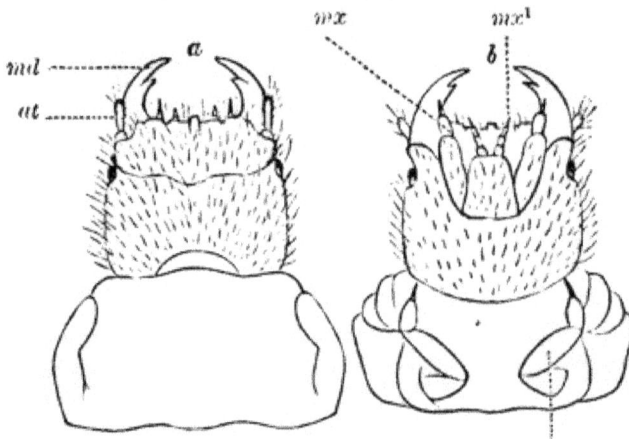

5. Mouth-parts of the Larva of a Beetle.

the true jaws, but the accessory jaws (maxillæ, Fig. 5, *a*, upper,
*b*, under side of the head of a young beetle; *at*, antennæ, *md*, mandible, *mx*, maxillæ, *mx¹*, labium) and the
feelers (palpi) attached to the maxillæ. and the under lip (labium) are of great service in enabling the
insect to detect its food both by the senses of touch
and smell. The maxillæ are in the fully grown beetle (Fig. 6) divided into three lobes, the outermost
forming the palpus, and the two others forming sharp teeth,
often provided with hairs and minute brushes for cleansing the
adjoining parts; these strong curved teeth are used in seizing

6. Maxilla of
a Beetle.

the food and placing it between the grinders, where it is crushed, prepared for digestion and swallowed.  Fig. 7 represents
the mouth parts
of the humble bee
(*b*, upper lip; *d*,
mandible; *c*, max-
illa; *f*, maxillary
palpus; *g*, tongue;
*ih*, labium and la-
bial palpi; *k*, eye.)

The alimentary
canal passes
through the middle
of the body, the
stomach forming
usually a simple
enlargement. Just
before the stomach
in certain insects,

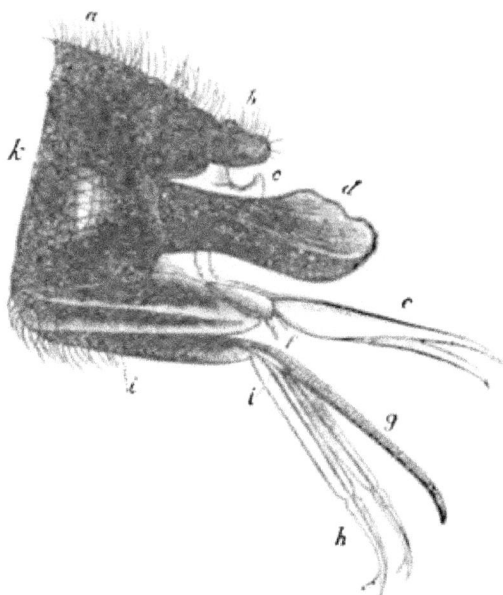

7. Mouth parts of a Humble Bee.

as the grasshopper, is a gizzard armed with rows of powerful
horny teeth for finely crushing grass.

Insects eat almost incredible quantities of food when young
and growing rapidly.  Mr. Trouvelot tells us in the " American
Naturalist" that the food taken by a single American Silk-
worm in fifty-six days is equal to eighty-six thousand times its
primitive weight!  On the other hand, after the insect has
finished its transformations, it either takes no food at all, as in
the May fly, or merely sips the honey of flowers, as in the butterfly, while the June beetle and many others like it eat the leaves
of trees, and the tiger and ground beetles feed voraciously on
other insects.

*How Insects Walk.*  In man and his allies, the vertebrates. the
process of walking is a most difficult and apparently dangerous
feat.  To describe the mechanics of walking, the wonderful

adaptation of the muscles and bones for the performance of this most ordinary action of life, would require a volume. The process is scarcely less complex in insects. Lyonnet found 3,993 muscles in a caterpillar, and while a large proportion belong to the internal organs, over a thousand assist in locomotion. Hence the muscular power of insects is enormous. A flea will leap two hundred times its own height, and certain large, solid beetles will move enormous weights as compared to the bulk of their bodies.

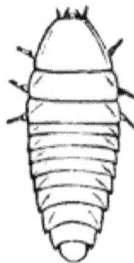

In walking, as seen in the accompanying figure (Fig. 8), three legs are thrown forward at a time, two on one side and one on the other.

Flies and many other insects can walk upside down, or on glass, as easily as on a level surface.

8. Larva of a beetle (Photuris). A fly's foot, as in most other insects, consists of five joints (tarsal joints), to the last one of which is appended a pair of stout claws, beneath which is a flat, soft, fleshy cushion or pad, split into two (sometimes three) flaps, beset on the under surface with fine hairs. A part of these hairs are swollen at the end, which is covered with "an elastic membranous expansion, capable of close contact with a highly polished surface, from which a minute quantity of a clear, transparent fluid is emitted when the fly is actively moving." (T. West.) These hairs are hence called holding, or tenent, hairs. With the aid of these, but mainly, as Mr. West insists, by the pressure of the atmosphere, a fly is enabled to adhere to perfectly smooth surfaces. His studies show the following curious facts. "That atmospheric pressure, if the area of the flaps be alone considered, is equal to just one-half the weight of a fly. If the area covered by the tenent hairs be added, an increase of pressure is gained, equal to about one-fourth the weight of a fly. This leaves one-fourth to be accounted for by slight viscidity of the fluid, by the action I have so often alluded to, which may be called 'grasping,' by molecular attraction, and,

doubtless, by other agents still more subtle, with which we have
at present scarcely any acquaintance."

*How Insects Fly.* Who of us, as remarked by an eminent orni-
thologist, can even now explain the long sustained, peculiar
flight of the hawk, or turkey buzzard, as it sails in the air without changing the position of its wings? and, we would add, the
somewhat similar flight of a butterfly? It is the poetry of motion, and a marvellous exhibition of grace and ease, combined
with a wonderful underlying strength and lightness of the parts
concerned in flight.

Before we give a partial account of the results obtained by
the delicate experiments of Professor Marey on the flight of
birds and insects, our readers should be reminded of the great
differences between an insect and a bird, remembering that the
former, is, in brief, a chitinous sac, so to speak, or rather a
series of three such spherical or elliptical sacs (the head, thorax
and abdomen); the outer walls of the body forming a solid but
light crust, to which are attached broad, membranous wings,
the wing being a sort of membranous bag stretched over a
framework of hollow tubes (the tracheæ), so disposed as to give
the greatest lightness and strength to the wing. The wings are
moved by powerful muscles of flight, filling up the cavity of the
thorax, just as the muscles are the largest about the thorax of
a bird. Moreover in the bodies of insects that fly (such as the
bee, cockchafer, and dragon fly), as distinguished from those
that creep exclusively, the air tubes (tracheæ) which ramify
into every part of the body, are dilated here and there, especially in the base of the abdomen, into large sacs, which are filled
with air when the insect is about to take flight, so that the
specific gravity of the body is greatly diminished. Indeed,
these air sacs, dilatable at will by the insect, may be compared
to the swimming bladder of fishes, which enables them to rise
and fall at will to different levels in the sea, thus effecting an
immense saving of the labor of swimming. In the birds, as

every body knows who has eaten a chicken, or attended the
dissection of a Thanksgiving turkey, the soft parts are external,
attached to the bony framework comprising the skeleton, the
wing bones being directly connected with the central back
bone; so that while these two sorts of animated flying machines
are so different in structure, they yet act in much the same manner when on the wing. The difference between them is clearly
stated by Marey, some of whose conclusions we now give almost
word for word.

The flight of butterflies and moths differs from that of birds
in the almost vertical direction of the stroke of their wings, and
in their faculty of sailing in the air without making any movements; though sometimes in the course they pursue they seem
to resemble birds in their flight.

The flight of insects and birds moreover differs in the form
of the trajectory in space; in the inclination of the plane in

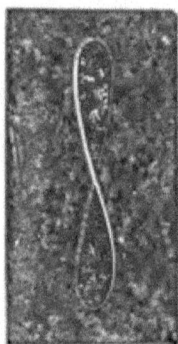

which the wings beat; in the role of each of
the two alternating (and in an inverse sense)
movements that the wings execute; as also in
the facility with which the air is decomposed
during these different movements. As the
wings of a fly are adorned with a brilliant
array of colors, we can follow the trajectory
or figure that each wing writes in the air. It
is of the form of a figure of eight (Fig. 9), first

9. Figure cut by an   discovered by Professor J. Bell Pettigrew of ·
insect's wing.        Edinburgh.

By an ingenious machine, specially devised for
the purpose, Professor Marey found that a bird's
wing moves in an ellipse, with a pointed summit
(Fig. 10). The insect beats the air in a distinctly
horizontal plane, but the bird in a vertical plane.

The wing of an insect is impervious to the air;
while the bird's wing resists the air only on its

10. Figure cut
by a bird's
wing.

under side. Hence, there are two sorts of effects; in the insect the up and down strokes are active; in the bird, the lowering of the wing is the only active period, though the return stroke seems to sustain the bird, the air acting on the wing. The bird's body is horizontal when the wing gives a downward stroke; but when the beat is upward, the bird is placed in an inclined plane like a winged projectile, and mounts up on the air by means of the inclined surfaces that it passively offers to the resistance of this fluid.

In an insect, an energetic movement is equally necessary to strike the air at both beats up and down. In the bird, on the contrary, one active beat only is necessary, the down beat. It creates at that time all the motive force that will be dispensed during the entire revolution of the wing. This difference is due to the difference in form of the wing. The difference between the two forms of flight is

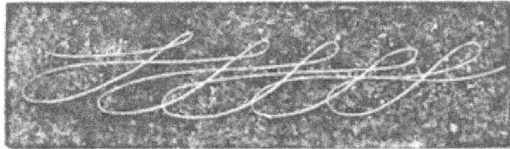

11. Trajectory of an insect's wing.

shown by an inspection of the two accompanying figures (11, 12). An insect's wing is small at the base and broad at the end. This breadth would be useless near the body, because at this point the wing does not move swiftly enough to strike the air effectively. The type of the insectean wing is designed, then, simply to strike the air. But in the bird the wing plays also a passive role, i. e.,

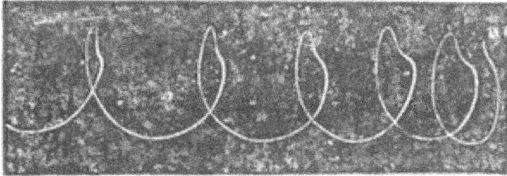

12. Trajectory of a bird's wing.

it receives the pressure of the air on its under side when the bird is projected rapidly onward by its acquired swiftness. In these conditions the whole animal is carried onward in space; all the points of its wing have the same velocity. The neighboring regions of the body are useful to press upon the air, which acts as on a paper kite. The base of the wing also, in the bird, is broad, and provided with feathers, which form a broad surface, on which the air presses with a force and method very efficacious in supporting the bird. Fig. 12 gives an idea of this disposition of the wing at the active and passive time in a bird.

The inner half of the wing is the passive part of the organ, while the external half, that which strikes the air, is the active part. A fly's wing makes 330 revolutions in a second, executing consequently 660 simple oscillations; it ought at each time to impress a lateral deviation of the body of the insect, and destroy the velocity that the preceding oscillation has given it in a contrary direction. So, that by this hypothesis the insect in its flight only utilizes fifty to one hundred parts (or one-half) of the resistance that the air furnishes it.

In the bird (Fig. 13), at the time of lowering the wings, the oblique plane which strikes the air, in decomposing the resistance, produces a vertical component which resists the weight of the body, and a horizontal component which imparts swiftness. The horizontal component is not lost, but is utilized during the rise of the wing, as in a paper kite when held in the

13. A bird on the wing.

air against the wind. Thus the bird utilizes seventy-five out of one hundred parts of the resistance that the air furnishes. The style of flight of birds is, therefore, theoretically superior to that of insects. As to the division of the muscular force between the resistance of the air and the mass of the body of the bird, we should compare the exertion made in walking on sand, for example, as compared with walking on marble. This is easy to measure. When a fish strikes the water with its tail to propel itself forward, it performs a double task; one part consists in pushing backwards a certain mass of water with a certain swiftness, and the other in pushing on the body in spite of the resistance of the surrounding fluid. This last portion of the task only is utilized. It would be greater if the tail of the fish encountered a solid object. Almost all the propelling agencies employed in navigation undergo this loss of labor, which depends on the mobility of the *point d' appui*. The bird is placed among conditions especially unfavorable.

*The Senses of Insects.* The eyes of insects are sometimes so large as to envelop the head like an Elizabethan ruffle, and the creature's head, as in the common house fly, seems all eyes. And this is almost literally the case, as the two great staring eyes that almost meet on the top of the head to form one, are made up of myriads of simple eyes. Each facet or simple eye is provided with a nerve filament which branches off from the main optic nerve, so that but one impression of the object perceived is conveyed to the brain; though it is taught by some that objects appear not only double but a thousand times multiplied. But we should remember that with our two eyes we see double only when the brain is diseased. Besides the large ordinary compound eyes, many insects possess small, simple eyes, like those of the spider. The great German anatomist, Johannes Müller, believed that the compound eyes were adapted for the perception of distant objects, while those nearer are seen by the simple eyes. But it may be objected to this view that the spiders, which have only simple eyes, apparently see both near and remote objects as well as insects.

The sense of touch is diffused all over the body. As in the hairs of the head and face of man, those of insects are delicate tactile organs; and on the antennæ and legs (insects depending on this sense rather than that of sight) these appendages are covered with exquisitely fine hairs. It is thought by some that the senses of hearing and smell are lodged in the antennæ, these organs thus combining the sense of feeling with those of hearing and smelling. And the researches of anatomists lend much probability to the assertion, since little pits just under the skin are found, and even sometimes provided with grains of sand in the so-called ear of the lobster, etc., corresponding to the ear bones of the higher animals, the pits being connected with nerves leading to the brain. We have detected similar pits in the under side of the palpi of the Perla. It seems not improbable that these are organs of smell, and placed in that part of the appendage nearest the mouth, so as to enable the insect to select its proper food by its odor. Similar organs exist on the caudal appendages of a kind of fly (Chrysopila), while the long, many-jointed caudal filaments of the cockroach are each provided with nearly a hundred of these little pits, which seem to be so many noses. Thus Lespès, a Swiss anatomist, in his remarks on the auditory sacs, which he says are found in the antennæ of nearly all insects, declares that as we have in insects compound eyes, so we have compound ears. We might add that in the abdominal appendage of the cockroach we have a compound nose, while in the feelers of the Perla, and the caudal appendage of the Chrysopila, the "nose" is simple. We might also refer here to Siebold's discovery of ears at the base of the abdomen of some, and in the forelegs of other kinds, of grasshoppers. Thus we need not be surprised at finding ears and noses scattered, as it were, sometimes almost wantonly over the bodies of insects (in many worms the eyes are found all over the body), while in man and his allies, from the monkey down to the fish, the ears and nose invariably retain the same relative place in the head.

*How Insects Grow.* When beginning our entomological studies no fact seemed more astonishing to our boyish mind than the thought that the little flies and midges were not the sons and daughters of the big ones. If every farmer and gardener knew this single fact it would be worth their while. The words *larva* and *pupa* will frequently occur in subsequent pages, and they should be explained. The caterpillar (Fig. 14, *a*) represents the earliest stage or babyhood of the butterfly, and it is called *larva*, from the Latin, meaning a mask, because it was thought by the ancients to mask the form of the adult butterfly.

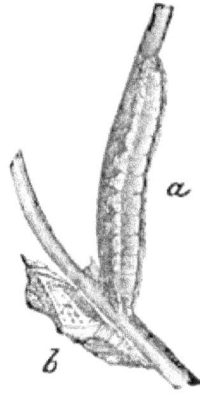

14. *a* Larva, *b* chrysalis of a butterfly.

When the caterpillar has ended its riotous life, for its appetite almost transforms its being into the very incarnation of gluttony, it suddenly, as if repenting of its former life as a *bon vivant*, seeks a solitary cell or hole where like a hermit it sits and leads apparently about as useless an existence. But meanwhile strange processes are going on beneath the skin; and after a few convulsive struggles the back splits open, and out wriggles the chrysalis, a gorgeous, mummy-like form, its body adorned with golden and silvery spots. Hence the word chrysalis (Fig. 14, *b*), from the Greek, meaning golden, while the Latin word *pupa*, meaning a baby or doll, is indicative of its youth. In this state it hangs suspended to a twig or other object; while the silk worm, and

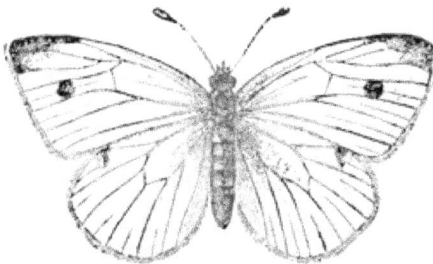

15. Imago or adult Butterfly.

others of its kind, previous to moulting, or casting their skins, spin a silken cocoon, which envelops and protects the chrysalis.

At the given time, and after the body of the adult has fully formed beneath the chrysalis skin, there is another moult, and the butterfly, with baggy, wet wings, creeps out. The body dries, the skin hardens, the wings expand, and in a few moments, sometimes an hour, the butterfly (Fig. 15) proudly sails aloft, the glory and pride of the insect world.

We shall see in the ensuing chapters how varied are the larvæ and pupæ of insects, and under what different guises insects live in their early stages.

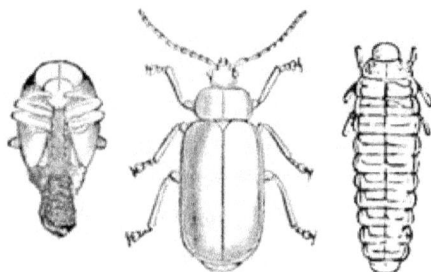

Larva, pupa, and adult of a Leaf Beetle (Galeruca).

# OUR COMMON INSECTS.

---

## CHAPTER I.

### THE HOME OF THE BEES.

THE history of the Honey bee, its wonderful instincts, its elaborate cells and complex economy, have engrossed the attention of the best observers, even from the time of Virgil, who sang of the Ligurian bee. The literature of the art of bee-keeping is already very extensive. Numerous bee journals and manuals of bee-keeping testify to the importance of this art, while able mathematicians have studied the mode of formation of the hexagonal cells,* and physiologists have investigated the intricate problems of the mode of generation and development of the bee itself.

In discussing these difficult questions, we must rise from the study of the simple to the complex, remembering that—

> "All nature widens upward. Evermore
> The simpler essence lower lies:
> More complex is more perfect—owning more
> Discourse, more widely wise,"

and not forget to study the humbler allies of the Honey bee. We shall, in observing the habits and homes of the wild bees, gain a clearer insight into the mysteries of the hive.

The great family of bees is divided into social and solitary species. The social kinds live in nests composed of numerous cells in which the young brood are reared. These cells vary in form from those which are quite regularly hexagonal, like those of the Hive bee, to those which are less regularly six-sided, as in the stingless bee of the tropics (Melipona), until in the Humble bee the cells are isolated and cylindrical in form.

---

* The cells are not perfectly hexagonal. See the studies on the formation of the cells of the bee, by Professor J. Wyman, in the Proceedings of the American Academy of Arts and Sciences, Boston, 1865; and the author's Guide to the Study of Insects, p 125.

Before speaking of the wild bees, let us briefly review the life of the Honey bee. The queen bee having wintered over with many workers, lays her eggs in the spring, first in the worker, and, at a later period, in the drone-cells. Early in the summer the workers construct the large, flask-shaped queen-cells, which are placed on the edge of the comb, and in these the queen larvæ are fed with rich and choice food. The old queen deserts the nest, forming a new colony. The new-born queen takes her marriage flight high in the air with a drone, and on her return undertakes the management of the hive, and the duty of laying eggs. When the supply of queens is exhausted, the workers destroy the drones. The first brood of workers live about six weeks in summer, and then give way to a new brood. The queens, according to Von Berlepsch, are known to live five years, and during their whole life lay more than a million eggs.

In the tropics, the Honey bee is replaced by the Meliponas and Trigonas. They are minute, stingless bees, which store up honey and live in colonies often of immense extent. The cells of Melipona are hexagonal, nearly approaching in regularity those of the Hive bee, while the honey cells are irregular, being much larger cavities, which hold about one-half as much honey as a cell of the Humble bee. "Gardner, in his travels, states that many species of Melipona build in the hollow trunks of trees, others in banks; some suspend their nests from the branches of trees, whilst one species constructs its nest of clay, it being of large size." (F. Smith.)

In a nest of the coal-black Trigona (Trigona carbonaria), from eastern Australia, Mr. F. Smith, of the British Museum, found from four hundred to five hundred dead workers, but no females. The combs were arranged precisely similar to those of the common wasp. The number of honey-pots which were placed at the foot of the nest was two hundred and fifty. Mr. Smith inclines to the opinion that the hive of Trigona contains several prolific females, as the great number of workers can only be thus explained, and M. Guérin found six females in a nest of the Tawny-footed Melipona (M. fulvipes).

At home, our nearest ally of the true Honey bee, is the Humble bee (Bombus), of which over forty species are known to inhabit North America.

The economy of the Humble bee is thus: the queen awakens in early spring from her winter's sleep under leaves or moss,

or in the last year's nest, and selects a nesting place, generally in an abandoned nest of a field-mouse, or beneath a stump
or sod, and "immediately," according to Mr. F. W. Putnam,*
"collects a small amount of pollen mixed with honey, and in
this deposits from seven to fourteen eggs, gradually adding to
the pollen mass until the first brood is hatched. She does not
wait, however, for one brood to be hatched before laying the
eggs of another, but, as soon as food enough has been collected,
she lays the eggs for a second. The eggs are laid, in contact
with each other, in one cavity
of the mass of pollen, with a
part of which they are slightly
covered. They are very soon
developed; in fact, the lines are
nowhere distinctly drawn between the egg and the larva,
the larva and pupa, and again
between the latter and the ima-         15. Cell and Eggs of Bombus.
go; a perfect series, showing this gradual transformation of
the young to the imago can be found in almost every nest.

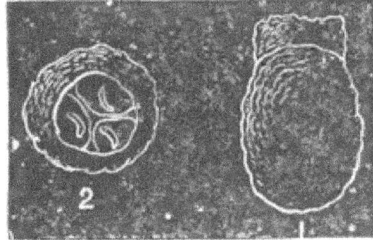

"As soon as the larvæ are capable of motion and commence
feeding, they eat the pollen by which they are surrounded, and,
gradually separating, push their way in various directions.
Eating as they move, and increasing in size quite rapidly, they
soon make large cavities in the pollen mass. When they have
attained their full size, they spin a silken wall about them, which
is strengthened by the old bees covering it with a thin layer of
wax, which soon becomes hard and tough, thus forming a cell
(Fig. 15, 1, cell containing a larva, on top of which (2) is a pollen mass containing three eggs). The larvæ now gradually at-

* Notes on the Habits of the Humble Bee (Proceedings of the Essex Institute, vol. iv, 1864, p. 101).

Mr. Angus also writes us as follows concerning the habits of the Wan-
dering Humble bee (Bombus vagans): "I have found the males plentiful
near our garden fence, within a hole such as would be made by a mouse.
They seem to be quite numerous. I was attracted to it by the noise they
were making in fanning at the opening. I counted at one time as many
as seven thus employed, and the sound could be heard several yards off.
Several males were at rest, but mostly on the wing, when they would make
a dash among the fanners, and all would scatter and play about. The
workers seem to be of a uniform size, and full as large as the males. I
think the object of the fanning was to introduce air into the nest, as is done
by the Honey bees."

tain the pupa stage, and remain inactive until their full development. They then cut their way out, and are ready to assume
their duties as workers, small females, males or queens.

"It is apparent that the irregular disposition of the cells is
due to their being constructed so peculiarly by the larvæ. After
the first brood, composed of workers, has come forth, the queen
bee devotes her time principally to her duties at home, the
workers supplying the colony with honey and pollen. As the
queen continues prolific, more workers are added, and the nest
is rapidly enlarged.

"About the middle of summer, eggs are deposited, which
produce both small females and males." . . . "All eggs laid
after the last of July produce the large females, or queens, and,
the males being still in the nest, it is presumed that the queens
are impregnated at this time, as on the approach of cold weather
all except the queens, of which there are several in each nest,
die."

While the Humble bee in some respects shows much less
instinct than the solitary bees mentioned below, it stands higher
in the series, however, from having workers, as well as males
and females, who provide food for the young. The labors of
the Mason bees, and their allies, terminate after the cell is once
constructed and filled with pollen. The eggs are then left to
hatch, and the young care for themselves, though the adult bee
shows greater skill in architecture than the Humble bee. It is
thus throughout nature. Many forms, comparatively low in the
scale of life, astonish us with certain characters or traits, reminding us of beings much superior, physically and intellectually.

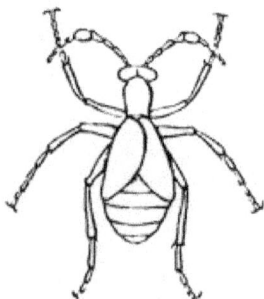

16. Meloë.

The lower forms constantly reach up
and in some way ally themselves with
creatures far more highly organized.
Thus the fish-like seal reminds us
strikingly of the dog, both in the form
of the head, in its docility and great intelligence when tamed, and even in its
bark and the movements of the head.

The parasites of the Humble bee are
numerous. Such are the species of
Apathus, which so closely resembles
the Humble bee itself, that it requires long study to distinguish
it readily. Its habits are not known, other than that it is found

in the nests of its host. It differs from the Humble bee in having no pollen-basket, showing that its larvæ must feed on the food stored up by their host, as it does not itself collect it. The mandibles also are not, like those of Bombus, trowel-shaped for architectural purposes, but acutely triangular, and are probably not used in building.

The caterpillars of various moths consume the honey and waxen cells; the two-winged flies, Volucella and Conops, and the larvæ of what is either an Anthomyia or Tachina-like fly, and several species of another genus of flies, Anthrax, together with several beetles, such as the Meloë (Fig. 16), Stylops (Fig. 17, male; 18b, female; a, position in the body of its host), and Antherophagus prey upon them.

The power of boring the most symmetrical tunnels in solid wood reaches its perfection in the large Virginian Carpenter bee

17. Male Stylops.

(Xylocopa Virginica, Fig. 19). This bee is as large as, and some allied exotic species are often considerably larger than, the Humble bee, but not clothed with such dense hairs. We have received from Mr. James Angus, of West Farms, N. Y., a piece of trellis from a grape vine, made of pine wood, containing the cells and young in various stages of growth, together with the larvæ and chrysalids of Anthrax sinuosa (Fig. 20), a species of fly parasitic on the larva. The maggot buries its head in the soft body of the young bee and feeds on its juices.

Mr. Angus thus writes us regarding its habits, under date of July 19: "I asked an intelligent and observing carpenter yesterday, if he knew how long it took the Xylocopa to bore her tunnel. He said he thought she bored about one-quarter of an inch a day. I don't think myself she bores more than one-half inch, if she does that. If I mistake not, it takes her about two

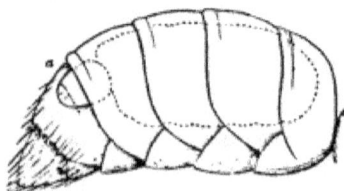

days to make her own length at the first start; but this being across the grain of the wood, may not be so easily done as the remainder, which runs parallel with it. She always follows the grain of the wood, with the exception of the entrance, which is about her own length. The tunnels run from one to one and a half feet in length. They generally run in opposite directions from the opening, and sometimes other galleries are run, one directly above the other, using the same opening. I think they only make new tunnels when old ones

18. Female Stylops.

are not to be found, and that the same tunnels are used for many years. Some of the old tunnels are very wide. I have found parts of them about an inch in diameter. I think this is caused by rasping off the sides to procure the necessary material for constructing their cells. The partitions are composed of wood raspings, and some sticky fluid, probably saliva, to make them adhere.

"The tunnels are sometimes taken possession of by other bees and wasps. I think when this is the case, the Xylocopa prefers making a new cell, to cleaning out the dirt and rubbish of the other species. I frequently find these bees remaining for a long time on the wing close to the opening, and bobbing their heads against the side, as if fanning air into the opening. I have seen them thus employed for twenty minutes. Whether one bee or more makes the tun-

19. Carpenter Bee.

nel, that is, whether they take turns in boring, I cannot at present say. In opening the cells (Fig. 21), more than one are generally found, even at this season. About two weeks ago, I found as many as seven, I think, in one." *

The hole is divided by partitions into cells about seven-tenths of an inch long. These partitions are constructed of the coarse dust or chippings made by the bee in eating out her cells, for our active little carpenter is provided with strong cutting jaws, moved by powerful muscles, and on her legs are stiff brushes of hair for cleaning out the tunnel as she descends into the heart of the solid wood. She must throw out the chips she bites off with her powerful mandibles from the sides of the burrow, by means of her hind legs, passing the load of chips backwards out of the cell with her fore limbs, which she uses as hands.

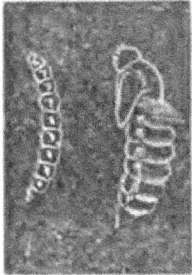

The partitions are built most elaborately of a single flattened band of chips, which is rolled up into a coil four layers deep. One side, forming the bottom of the cell, is concave, being beaten down and smoothed off by the bee. The other side of the partition, forming the top of the cell, is flat and rough.

20. Larva and Pupa of Anthrax.

At the time of opening the burrow, July 8th, the cells contained nearly full-grown larvæ, with some half developed. They were feeding on the masses of pollen, which were as large as a thick kidney bean, and occupied nearly half the cell. The larva (Fig. 21) resemble those of the Humble bee, but are slenderer, tapering more rapidly towards each end of the body.

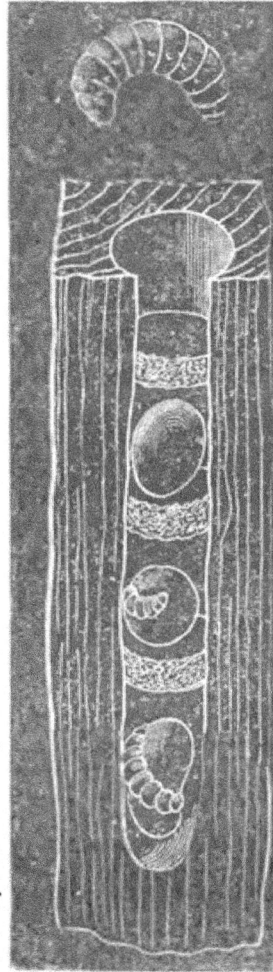

21. Nest of Carpenter Bee.

* " Since writing the above I have opened one of the new holes of Xylocopa, which was commenced between three and four weeks ago, in a pine

The habits and structure of the little green Ceratina ally it closely with Xylocopa. This pretty bee, named Ceratina dupla by Mr. Say, tunnels out the stems of the elder or blackberry, syringa, or any pithy shrub, excavating them often to a depth of six or seven inches. She makes the walls just wide enough to admit her body, and of a depth capable of holding three or · four, often five or six cells (Fig. 22). The finely built cells, with their delicate silken walls, are cylindrical and nearly square at each end, though the free end of the last cell is rounded off. They are four and a half tenths of an inch long, and a little over one-third as broad. The bee places them at nearly equal distances apart, the slight interval between them being filled in with dirt.

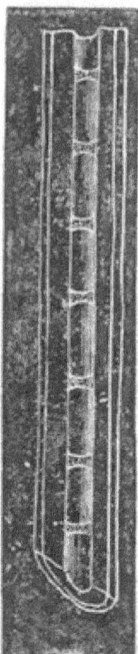

Dr. T. W. Harris states that May 15, 1832, one female laid its eggs in the hollow of an aster stalk. Three perfect insects were disclosed from it July 28th. The observations of Mr. Angus, who saw some bees making their cells May 18th, also confirm this account. The history of our little upholsterer is thus cleared up. Late in the spring she builds her cells, fills them with pollen, and lays one or more eggs upon each mass. Thus in about two months the insect completes its transformations; within this period passing through the egg, the larva and chrysalid states, and then, as a bee, living a few days more, if a

22. Nest of Ceratina.

slit used in the staging of the greenhouse. The dimensions were as follows:—Opening fully 3-8 wide; depth 7-16; whole length of tunnel 6 5-16 inches. The tunnel branched both ways from the hole. One end, from opening, was 2 5-8, containing three cells, two with larva and pollen, the third empty. The other side of the opening, or the rest of the tunnel, was empty, with the exception of the old bee (only one) at work. I think this was the work of one bee, and, as near as I can judge, about twenty-five days' work. Width of tunnel inside at widest 9-16 inch.

I have just found a Xylocopa bobbing at one of the holes, and in order to ascertain the depth of the tunnel, and to see whether there were any others in them, I sounded with a pliable rod, and found others in one side, at a depth of five and one half inches; the other side was four inches deep without bees. The morning was cool, so that the object in bobbing could not have been to introduce fresh currents of air, but must have had some relation to those inside. Their legs on such occasions are, as I have noticed, loaded with pollen."

male; or if a female, living through the winter. Her life thus spans one year.

The larva (Fig. 23) is longer than that of Megachile, and compared with that of Xylocopa, the different segments are much more convex, giving a serrate outline to the back of the worm. The pupa, or chrysalis, we have found in the cells the last of July. It is white, and three-tenths of an inch long. It

23. Larva of Ceratina.

differs from that of the Leaf-cutter bee in having four spines on the end of the body.

In none of the wild bees are the cells constructed with more nicety than those of our little Ceratina. She bores out with her jaws a long deep well just the size of her body, and then stretches a thin, delicate cloth of silk drawn tight as a drum-head across each end of her chambers, which she then fills with a mixture of pollen and honey.

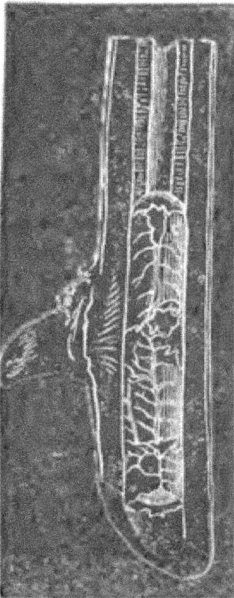

24. Nest of Tailor Bee.

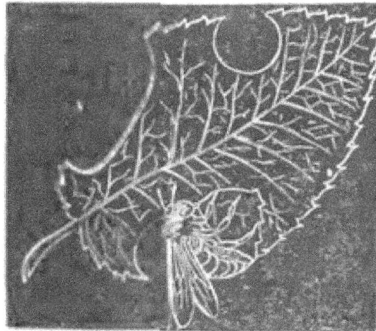

25. Tailor Bee.

Her young are not, in this supposed retreat, entirely free from danger. The most invidious foes enter and attack the brood. Three species of Ichneumon flies, two of which belong to the Chalcid family, lay their eggs within the body of the larva, and emerge from the dried larva and pupa skins of the bee, often in great numbers. The smallest parasite, belonging to the genus Anthophorabia, so called from being first known as a parasite on another bee (Anthophora), is a minute species found also abundantly in the tight cells of the Leaf-cutter bee.

3

The interesting habits of the Leaf-cutting, or Tailor bee
(Megachile), have always attracted attention. This bee is a
stout, thick-bodied insect, with a large, square head, stout,
sharp, scissors-like jaws, and with a thick mass of stout, dense
hairs on the under side of the tail for carrying pollen, as she is
not provided with the pollen-basket of the Honey and Humble
bees.

The Megachile lays its eggs in burrows in the stems of the
elder (Fig. 24), which we have received from Mr. James Angus;
we have also found them in the hollows of the locust tree.  Mr.
F. W. Putnam thus speaks of the economy of M. centuncularis,
our most common species.  "My attention was first called, on
the 26th of June, to a female busily engaged in bringing pieces
of leaf to her cells, which she was building under a board, on
the roof of the piazza, directly under my window.  Nearly the
whole morning was occupied by the bee in bringing pieces of
leaf from a rose bush growing about ten yards from her cells,
returning at intervals of a half minute to a minute with the
pieces, which she carried in such a manner as not to impede
her steps when she alighted near her hole."  When the Leaf-
cutter bee wishes to cut out a piece of a leaf (Fig. 25) she alights
upon the leaf, and in a few seconds swiftly runs her scissors-
like jaws around through it, bearing off the piece in her hind
legs.  "About noon she had probably completed the cell, upon
which she had been engaged, as, during the afternoon, she was
occupied in bringing pollen, preparatory to laying her single
egg in the cell.  For about twenty days the bee continued at
work, building new cells and supplying them with pollen. . . .
On the 28th of July, upon removing the board, it was found that
the bee had made thirty cells, arranged in nine rows of unequal
length, some being slightly curved to adapt them to the space
under the board.  The longest row contained six cells, and was
two and three-quarters inches in length; the whole leaf structure being equal to a length of fifteen inches.  Upon making an
estimate of the pieces of leaf in this structure, it was ascertained
that there must have been at least a thousand pieces used.  In
addition to the labor of making the cells, this bee, unassisted
in all her duties, had to collect the requisite amount of pollen
(and honey?) for each cell, and lay her eggs therein, when completed.  Upon carefully cutting out a portion of one of the cells,
a full-grown larva was seen engaged in spinning a slight silken

cocoon about the walls of its prison, which were quite hard and smooth on the inside, probably owing to the movements of the larva, and the consequent pressing of the sticky particles to the walls. In a short time the opening made was closed over by a very thin silken web. The cells, measured on the inside of the hard walls, were .35 of an inch in length, and .15 in diameter. The natural attitude of the larva is somewhat curved in its cell, but if straightened, it just equals the inside length of the cell. On the 31st of July, two female bees came out, having cut their way through the sides of their cells." In three other cells " several hundred minute Ichneumons (Anthophorabia megachilis) were seen, which came forth as soon as the cells were opened."

The habits of the little blue or green Mason bees (Osmia) are quite varied. They construct their cells in the stems of plants, and in rotten posts and trees, or, like Andrena, they burrow in sunny banks. A European species selects snail shells for its nest, wherein it builds its earthen cells, while other species nidificate under stones. Curtis found two hundred and thirty cocoons of a British species (Osmia parietina), placed on the under side of a flat stone, of which one-third were empty. Of the remainder, the most appeared between March and June, males appearing first; thirty-five more bees were developed the following spring. Thus there were three successive broods,

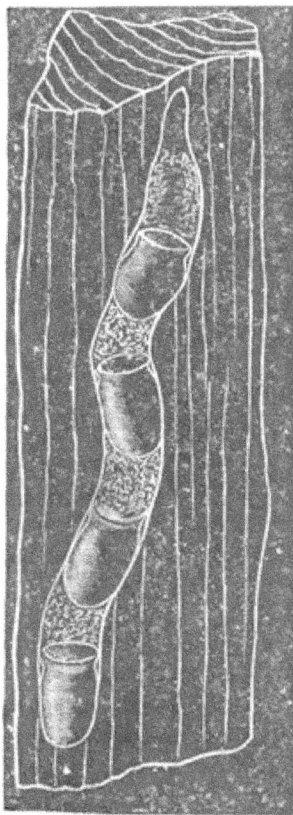

26. Nest of Osmia.

for three succeeding years, so that these bees lived three years before arriving at maturity. This may partly account for *insect years*, which are like "apple years," seasons when bees and wasps, as well as other insects, abound in unusual numbers.

Mr. G. R. Waterhouse, in the Transactions of the Entomological Society of London, for 1864, states that the cells of Osmia leucomelana "are formed of mud, and each cell is built separately. The female bee, having deposited a small pellet of mud in a sheltered spot between some tufts of grass, immediately begins to excavate a small cavity in its upper surface, scraping the mud away from the centre towards the margin by means of her jaws. A small, shallow mud-cup is thus produced. It is rough and uneven on the outer surface, but beautifully smooth on the inner. On witnessing thus much of the work performed, I was struck with three points: first, the rapidity with which the insect worked; secondly, the tenacity with which she kept her original position whilst excavating; and thirdly, her constantly going over work which had apparently been completed. . . . The lid is excavated and rendered concave on its outer or upper surface, and is convex and rough on its inner surface; and, in fact, is a simple repetition of the first-formed portion of the cell, a part of a hollow sphere."

The largest species of Osmia known to us is a very dark-blue species (O. lignivora). We are indebted to a lady for specimens of the bees with their cells, which had been excavated in the interior of a maple tree several inches from the bark. The bee had industriously tunnelled out this elaborate burrow (Fig. 26), and, in this respect, resembled the habits of the Carpenter bee more closely than any other species of its genus.

The tunnel was over three inches long, and about three-tenths of an inch wide. It contracted a little in width between the cell, showing that the bee worked intelligently, and wasted no more of her energies than was absolutely necessary. The burrow contained five cells, each half an inch long, being rather short and broad, with the hinder end rounded, while the opposite end, next to the one adjoining, is cut off squarely. The cell is somewhat jug-shaped, owing to a slight constriction just behind the mouth. The material of which the cell is composed is stout, silken, parchment-like, and very smooth within. The interstices between the cells are filled in with rather coarse chippings made by the bee.

The bee cut its way out of the cells in March, and lived for a month afterwards on a diet of honey and water. It eagerly lapped up the drops of water supplied by its keeper, to whom it soon grew accustomed, and seemed to recognize.

Our smallest and most abundant species is the little green Osmia simillima. It builds its little oval, somewhat urn-shaped cells against the roof of the large deserted galls of the oak-gall fly (Diplolepis confluentus), placing them, in this instance eleven in number, in two irregular rows, from which the mature bees issue through a hole in the gall (Fig. 27, with two separate cells). The earthen cells, containing the tough dense cocoons, were arranged irregularly so as to fit the concave vault of the larger gall, which was about two inches in diameter. On emerging from the cell the Osmia cuts out with its powerful jaws an ovate lid, nearly as large as one side of the cell.

In the Harris collection are the cells and specimens of Osmia pacifica, the peaceful Osmia, which, according to the manuscript

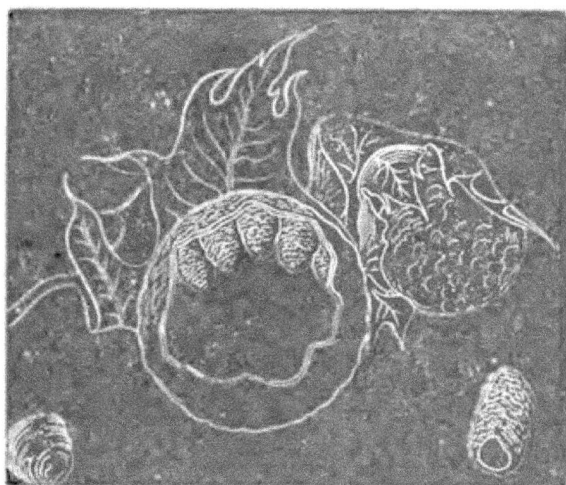

27. Nest of Osmia in a gall.

notes of Dr. Harris, is found in the perfect state in earthen cells beneath stones. The cell is oval cylindrical, a little contracted as usual with those of all the species of the genus, thus forming an urn-shaped cell. It is half an inch long, and nearly three-tenths of an inch wide, while the cocoon, which is rather thin, is three-tenths of an inch long. We are not acquainted with the habits of the larva and pupa in this country, but Mr. F. Smith states that the larva of the English species hatches in eight days after the eggs are laid, feeds ten to twelve days, when it becomes full-grown, then spins a thin silken covering, and remains in an inactive state until the following spring, when it completes its transformations.

In the economy of our wild bees we see the manifestation of a wonderful instinct, as well as the exhibition of a *limited reason*. We can scarcely deny to animals a kind of reason which apparently differs *only in degree* from that of man. Each species works in a sphere limited by physical laws, but within that sphere it is a free agent. They have enough of instinct and reason to direct their lives, and to enable them to act their part in carrying out the plan of creation.

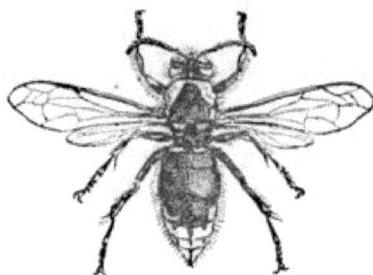

Paper Wasp.

# CHAPTER II.

[*Concluded.*]

WHILE the Andrena and Halictus bees, whose habits we now describe, are closely allied in form to the Hive bee, socially they are the "mud-sills" of bee society, ranking among the lowest forms of the family of bees. Their burrowing habits ally them with the ants, from whose nests their own burrows can scarcely be distinguished. Their economy does not seem to demand the exercise of so much of a true reasoning power and pliable instinct as characterizes bees, such as the Honey and Humble bee, which possess a high architectural skill. Moreover they are not social; they have no part in rearing and caring for their young, a fact that lends so much interest to the history of the Hive and Humble bee. In this respect they are far below the wasps, a family belonging next below in the system of Nature.

A glance at the drawing (Fig. 28), of a burrow, with its side galleries, of the Andrena vicina, reveals the economy of one of our most common forms. Quite early in spring, when the sun and vernal breezes have dried up the soil, and the fields exchange their rusty hues for the rich green verdure of May, our Andrena, tired of its idle life among the blossoms of the willow, the wild cherry, and garden flowers, suddenly becomes remarkably industrious, and wields its spade-like jaws and busy feet with a strange and unwonted energy. Choosing some sunny, warm, grassy bank (these nests were observed in the "great pasture" of Salem), not always with a southern exposure however, the female sinks her deep well through the sod from six inches to a foot into the sandy soil beneath. She goes to work literally tooth and nail. Reasoning from observations

(31)

Fig. 28.

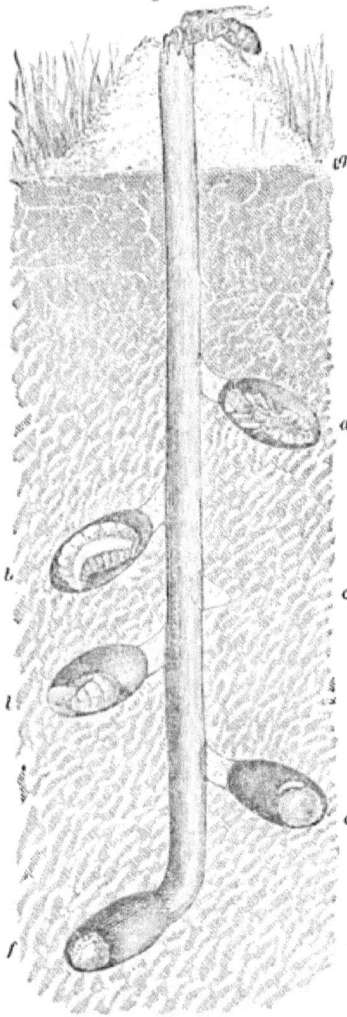

Nest (natural size) of Andrena vicina, showing the main burrow, and the cells leading from it; the oldest cell containing the pupa (*a*) is situated nearest the surface, while those containing the larva (*b*) lie between the pupa and the cell (*e*) containing the pollen mass and egg resting upon it. The most recent cell (*f*) is the deepest down, and contains a freshly deposited pollen mass. At *e* is the beginning of a cell; *g*, level of the ground.

made on several species of wasps, and also from studying the structure of her jaws and legs, it is evident that she digs in and loosens the soil with her powerful jaws, and then throws out the dirt with her legs. She uses her fore legs like hands, to pass the load of dirt to her hind legs, and then runs backward out of her hole to dump it down behind her. Mr. Emerton tells me that he never saw a bee in the act of digging but once, and then she left off after a few strokes. He also says, "they are harmless and inoffensive. On several occasions I have lain on the grass near their holes for hours, but not one attempted to sting me; and when taken between the fingers, they make but feeble resistance."

To enter somewhat into detail, we gather from the observations of Mr. Emerton (who has carefully watched the habits of these bees through several seasons) the following account of the economy of this bee: On the 4th of May the bees were seen digging their holes, most of which were already two inches deep, and one, six inches. The mounds of earth were so small as to be hardly noticed. At this time an Oil beetle was seen prowling about the holes. The presence of this dire foe of Andrena at this time, it will be seen in a succeeding chapter on the enemies of the bees, is quite significant. By the 15th of May, hundreds of Andrena holes were found in various parts of the pasture, and at one place, in a previous season, there were about two hundred found placed within a small area. One cell was dug up, but it contained no pollen. Four days later, several Andrenas were noticed resting from their toil at the opening of their burrows. On the 28th of May, in unearthing six holes, eight cells were found to contain pollen, and in two of them a small larva. The pellets of pollen are about the size of a small pea. They are hard and round at first, before the young has hatched, but as the larva grows, the mass becomes softer and more pasty, so that the larva buries its head in the mass, and greedily sucks it in. When is the pollen gathered by the bee and kneaded into the pellet-like mass? On July 4th, a cell was opened in which was a bee busily engaged preparing the pollen, which was loosely and irregularly piled up, while there was a larva in an adjoining cell nearly half an inch long. It would seem, then, that the bee comes in from the fields laden with her stores of pollen, which she elaborates into bee bread within her cell.

When the bee returns to her cell she does not directly fly towards the entrance, since, as was noticed in a particular instance, she flew about for a long time in all directions without any apparent aim, until she finally settled near the hole, and walked into her subterranean retreat. On a rainy day, May 24th, our friend visited the colony, but found no bees flying about the holes. The little hillocks had been beaten down by the pitiless raindrops, and all traces of their industry effaced. On digging down, several bees were found, indicating that on rainy days they seek the shelter of their holes, and do not take refuge under leaves of the plants they frequent.

On the 29th of June, six full-grown larvæ were exhumed, and one, about half grown. On the 20th of July, the colony seemed well organized, as, on laying open a burrow at the depth of six inches, he began to find cells. The upper ones, to the number of a dozen, were deserted and filled with earth and grass roots, and had evidently been built and used during the previous year. Below these were eight cells placed around the main vertical gallery, reaching down to the depth of thirteen inches, and all containing nearly full-grown larvæ of the bees, or else those of some parasitic bee (Nomada) which had devoured the food prepared for the young Andrena.

About the first of August the larva transforms to a pupa or chrysalis, as at this time two pupæ were found in cells a foot beneath the surface. As shown in the cut, those cells situated lowest down seem to be the last to have been made, while the eggs laid in the highest are the first to hatch, and the larvæ disclosed from them, the first to change to pupæ. Four days later the pupæ of Cuckoo bees (Nomada) were found in the cells. No Andrenas were seen flying about at this time.

On the 24th of August, to be still very circumstantial in our narrative though at the risk of being tedious, three burrows were unearthed, and in them three fully formed bees were found nearly ready to leave their cells, and in addition several pupæ. In some other cells there were three of the parasitic Nomada also nearly ready to come out, which seemed to be identical with some bees noticed playing very innocently about the holes early in the summer.

On the last day of August, very few of the holes were open. A number of Oil beetles were strolling suspiciously about in the neighborhood, and some little black Ichneumon flies were seen running about among the holes.

During midsummer the holes were found closed night and day by clods of earth.

The burrow is sunken perpendicularly, with short passages leading to the cells, which are slightly inclined downwards and outwards from the main gallery. The walls of the gallery are rough, but the cells are lined with a mucous-like secretion, which, on hardening, looks like the glazing of earthenware. This glazing is quite hard, and breaks up into angular pieces. It is evidently the work of the bee herself, and is not secreted and laid on by the larva. The diameter of the interior of the cell is about one-quarter of an inch, contracting a little at the mouth. When the cell is taken out, the dirt adheres for a line in thickness, so that it is of the size and form of an acorn.

The larva of Andrena (Fig. 29) is soft and fleshy, like that of the Honey bee. Its body is flattened, bulging out prominently at the sides, and tapering more rapidly than usual towards each end of the body. The skin is very thin, so that along the back the heart or dorsal vessel may be distinctly seen, pulsating about sixty times a minute.

Our cut (Fig. 28, a) also represents the pupa, or chrysalis, as seen lying in its cell. The limbs are folded close to the body in the most compact way possible. On the head of the semi-pupa, *i. e.*, a transition state between the larva and pupa, there are two prominent tubercles situated behind the simple eyes, or ocelli; these are deciduous organs, apparently aiding the insect in moving about its cell. They disappear in the mature pupa.

To those accustomed to rearing butterflies, and seeing the chrysalis at once assuming its perfected shape, after the caterpillar skin is thrown off, it may seem strange to hear one speak of a "half-pupa," and of stages intermediate between the larva and pupa. But the external changes of form, though rapidly passed through, consisting apparently of a mere sloughing off of the outer skin, are yet preceded by slow and very gradual alterations of tissues, resulting from the growth of cells. An inner layer of the larva-skin separates from the outer, and, by changes in the form of the muscles, is drawn into different positions, such as is assumed by the pupa, which thus lies concealed beneath the larva-skin. But a slight alteration is made in the general form of the larva, consisting mostly of an enlargement of the thoracic segments, which is often overlooked, even by the special student, though of great interest to the philosophic naturalist.

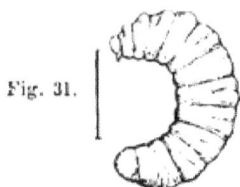

Fig. 31.

Fig. 29.

From Mr. Emerton's observations we should judge that the pupa state lasted from three to four weeks, as the larvæ began to transform the first of August, and appeared during the last week of the same month as perfect bees.

Fig. 30.

The Andrena is seen as late as the first week in September, and again early in April, about the flowers of the willow. It is one of the largest of its genus and a common species.

Fig. 31. Larva of Halic-tus parallelus.
Fig. 29. Larva of Andre-na vicina.
Fig. 30. Pupa of Halic-tus parallelus seen from beneath.

Having, in a very fragmentary way, sketched the life history of our Andrena, and had some glimpses of its subterranean life, let us now compare with it another genus of solitary bee

(Halictus), quite closely allied in all respects, though a little lower in the scale.

The Halictus parallelus excavates cells almost exactly like those of Andrena; but since the bee is smaller, the holes are smaller, though as deep. Mr. Emerton found one nest in a path a foot in depth. Another nest, discovered September 9th, was about six inches deep. The cells are in form like those of Andrena, and like them, are glazed within. The egg is rather slenderer and much curved; in form it is long, cylindrical, obtuse at one end, and much smaller at the other. The larva (Fig. 31) is longer and slenderer, being quite different from the rather broad and flattened larva of Andrena. The body is rather thick behind, but in front tapers slowly towards the head, which is of moderate size. Its body is somewhat tuberculated, the tubercle aiding the grub in moving about its cell. Its length is nearly one-half (.40) of an inch. On the pupa are four quite distinct conical tubercles forming a transverse line just in front of the ocelli; and there are also two larger, longer tubercles, on the outer side of each of which, an ocellus is situated. Figure 30 represents the pupa seen from beneath.

Search was made on July 16th, where the ground was hard as stone for six inches in depth, below which the soil was soft and fine, and over twenty cells were dug out. "The upper cells contained nearly mature pupæ, and the lower ones, larvæ of various sizes, the smallest being hardly distinguishable by the naked eye. Each of these small larvæ was in a cell by itself, and situated upon a lump of pollen, which was the size and shape of a pea, and was found to lessen· in size as the larva grew larger. These young were probably the offspring of several females, as four mature bees were found in the hole." The larva of an English species hatches in ten days after the eggs are laid.

Another brood of bees appeared the middle of September, as on the ninth of that month (1864) Mr. Emerton found several holes of the same species of bee, made in a hard gravel road near the turnpike. When opened, they were found to contain several bees with their young. September 2d, of this year, the same kind of bee was found in holes, and just ready to leave the cell. It is probable that these bees winter over.

We have incidentally noticed the presence in the nests of Andrena and Halictus of a stranger bee, clad in gay, fantastic hues, which lives a parasitic life on its hosts. This parasitism does not go far enough to cause the death of the host, since we find the young of the parasitic Cuckoo bee, in cells containing the young of the former.

Mr. F. Smith, in his "Catalogue of British Bees," says of this genus: "No one appears to know anything beyond the mere fact of their entering the burrows of Andrenidæ and Apidæ, except that they are found in the cells of the working bees in their perfect condition: it is most probable that they deposit their eggs on the provision laid up by the working bee, that they close up the cell, and that the working bee, finding an egg deposited, commences a fresh cell for her own progeny."

He has, however, found two specimens of Nomada sexfasciata in the cells of the long-horned bee, Eucera longicornis. He also states, that while some species are constant in their attacks on certain Halicti and Andrenæ, others attack different species of these genera indiscriminately. In like manner another Cuckoo bee (Cœlioxys) is parasitic on Megachile and Saropoda; Stelis is a parasite on Osmia, the Mason bee: and Melecta infests the cells of Anthophora.

The observations of Mr. Emerton enable us still further to clear up the history of this obscure visitor. He found both the larva and pupa, as well as the perfect bee, in the cells of both genera; so that either both kinds of bee, when hatched from eggs laid in the same cell, feed on the same pollen mass, which therefore barely suffices for the nourishment of both; or the hostess, discovering the strange egg laid, cuckoo-like, in her own nest, has the forethought to deposit another ball of pollen to secure the safety of her young.

Is such an act the operation of a blind instinct? Does it not rather ally our little bee with those higher animals which undoubtedly possess a reasoning power? Its *instinct* teaches it to build cells, and prepare its pollen mass, and lay an egg thereon. Its *reason* enables it, in such an instance as this, when the life of the brood is threatened, to guard against any such danger by means to which it does not habitually resort. This instance is paralleled by the case of our common summer Yellow bird, which, on finding an egg of the Cow bunting in its nest, often builds a new nest above it, to the certain destruction of the unwelcome egg in the nest beneath.

In the structure of the bee, and in all its stages of growth,

4

our parasite seems lower in the zoölogical scale than its host.
It is structurally a degraded form of Working-bee, and its position socially is unenviable.   It is lazy, not having the provi-
dent habits of the Working-bees; it aids not in the least, so far
as we know, the cross-fertilization of plants,—one great office
in the economy of nature which most bees perform,—since it is
not a pollen-gatherer, but on the contrary is seemingly a drag
and hinderance to the course of nature.   But yet nature kindly,
and as if by a special interposition, provides for its main-
tenance, and the humble naturalist can only exclaim, " God is
great, and his ways mysterious," and go on studying and collecting facts, leaving to his successors the more difficult task,
but greater joy of discovering the cause and reason of things
that are but a puzzle to the philosophers of this day.

The larva of Nomada may be known from those of its host, by
its slenderer body and smaller head, while the body is smoother
and more cylindrical.   Both sexes of Nomada imbricata and
N. pulchella were found by Mr. Emerton, the former in both the
Andrena and Halictus nests, and both were found in a single
Andrena nest.

Wood Wasp.

# CHAPTER III.

VERY few bee-keepers are probably aware how many insect parasites infest the Honey bee. In our own literature we hear almost nothing of this subject, but in Europe much has been written on bee parasites. From Dr. Edward Assmuss' little work on the "Parasites of the Honey Bee," we glean some of the facts now presented, and which cannot fail to interest the general reader as well as the owner of bees.

The study of the habits of animal parasites has of late gained much attention among naturalists, and both the honey and wild bees afford good examples of the singular relation between the host and the parasites which live upon it. Among insects generally, there are certain species which devour the contents of the egg of the victim. Others, and this is the most common mode of parasitism, attack the insect in its larva state; others, in the pupa state, and still others in the perfect, or imago state. Dr. Leidy has shown that the wood-devouring species of beetle, Passalus cornutus, and some Myriopods, or "thousand legs," are, in some cases, tenanted by myriads of microscopic plants and worms which luxuriate in the alimentary canal, while the "caterpillar-fungus" attacks sickly caterpillars, filling out their bodies, and sending out shoots into the air, so that the insect looks as if transformed into a vegetable.

The Ichneumon flies, of which there are undoubtedly several thousand species in this country, are the most common insect parasites. Next to these are the different species of Tachina and its allied genera. These, like Ichneumons, live in the bodies of their hosts, consuming the fatty parts, and finishing their

(39)

transformations just as the exhausted host is ready to die, issue from their bodies as flies, closely resembling the common house-fly.

A small fly has been found in Europe to be the most formid-able foe of the hive bee, sometimes producing the well-known disease called " foul-brood," which is analogous to the typhus fever of man.

This fly, belonging to the genus Phora (Fig. 32, Phora incras-sata; *a*, larva; *b*, puparium; *c*, another species from Mammoth

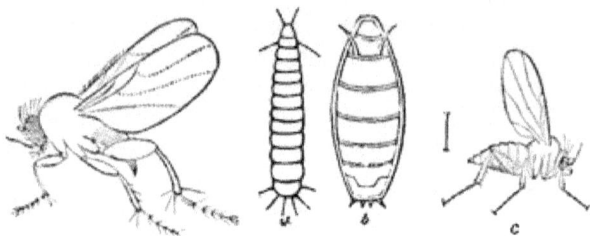

32. Phora and its Young.

Cave), is a small insect about a line and a half long, and found in Europe during the summer and autumn flying slowly about flowers and windows, and in the vicinity of beehives. Its white, transparent larva is cylindrical, a little pointed before, but broader behind. The head is small and rounded, with short, three-jointed antennæ, and at the posterior end of the body are several slender spines. The puparium, or pupa case, inclosing the delicate chrysalis, is oval, consisting of eight segments, flattened above, with two large spines near the head, and four on the extremity of the body.

When impelled by instinct to provide for the continuance of its species, the Phora enters the beehive and gains admission to a cell, when it bores with its ovipositor through the skin of the bee larva, laying its long oval egg in a horizontal position just under the skin. The embryo of the Phora is already well developed, so that in three hours after the egg is inserted in the body of its unsuspecting and helpless host, the embryo is nearly ready to hatch. In about two hours more it actually breaks off the larger end of the egg-shell and at once begins to eat the fatty tissues of its victim, its posterior half still remaining in the shell. In an hour more, it leaves the egg entirely and buries itself completely in the fatty portion of the young bee.

The maggot moults three times. In twelve hours after the last moult it turns around with its head towards the posterior end of the body of its host, and in another twelve hours, having become full-fed, it bores through the skin of the young, eats its way through the brood-covering of the cell and falls to the bottom of the hive, where it changes to a pupa in the dust and dirt, or else creeps out of the door and transforms in the earth. Twelve days after, the fly appears.

The young bee, emaciated and enfeebled by the attacks of its ravenous parasite, dies, and its decaying body fills the bottom of the cell with a slimy, foul-smelling mass, called "foul-brood." This gives rise to a miasma which poisons the neighboring brood, until the contagion (for the disease is analogous to typhus, jail or ship-fever) spreads through the whole hive, unless promptly checked by removing the cause and thoroughly cleansing the hive.

Foul-brood sometimes attacks our American hives, and, though the cause may not be known, yet from the hints given above we hope to have the history of our species of Phora cleared up, should our disease be found to be sometimes due to the attacks of such a parasitic fly.

We figure the Bee louse of Europe (Fig. 33 b, Braula cæca), which is a singular wingless spider-like fly, allied to the wingless Sheep tick (Melophagus), the wingless Bat tick (Nycteribia)

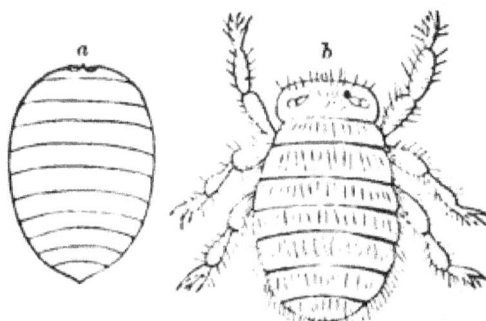

33. Bee Louse and Larva.

and the winged Horse fly (Hippobosca). The head is very large, without eyes or ocelli (simple eyes), while the ovate hind-body consists of five segments, and is covered with stiff hairs. It is one-half to two-thirds of a line long. This spider fly is "pupiparous," that is, the young, of which only a very few are produced, is not born until it has assumed the pupa state or is just about to do so. The larva (Fig. 33 a) is oval, eleven-jointed, and white in color. The very day it is hatched, it sheds its skin and changes to an oval puparium of a dark brown color.

Its habits resemble those of the flea. Indeed, should we compress its body strongly, it would bear a striking resemblance to that insect. It is evidently a connecting link between the flea, and the two winged flies. Like the former it lives on the body of its host, and obtains its food by plunging its stout beak into the bee and sucking its blood.

It has not been noticed in this country, but is liable to be imported on the bodies of Italian bees. Generally, one or two of the Braulas may, on close examination, be detected on the body of the bee; sometimes the poor bees are loaded down by as many as a hundred of these hungry blood-suckers. Assmuss recommends rubbing them off with a feather, as the bee goes in and out of the door of its hive.

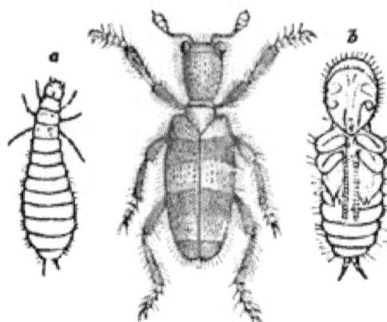

34. Hive Trichodes.

Among the beetles are a few forms occasionally found in bees' nests and also parasitic on the body of the bee. Trichodes apiarius (Fig. 34, a, larva; b, pupa, front view) has long been known in Europe to attack the young bees. In its perfect, or beetle state it is found on flowers, like our Trichodes Nuttallii, which is commonly found on the Spiræa in August, and which may yet prove to enter our beehives. The larva devours the brood, but with the modern hive its ravages may be readily detected.

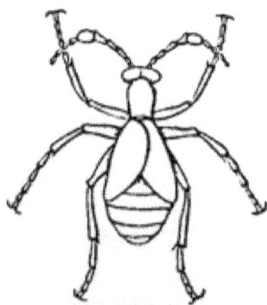

35. Meloë.

The Oil beetle, Meloë angusticollis (Fig. 35, male, differing from the female by having the antennæ as if twisted into a knot; Fig. 36, the active larva found on the body of the bee), is a large dark blue insect found crawling in the grass in the vicinity of the nests of Andrena, Halictus, and other wild bees in May, and again in August and September.

The eggs are laid in a mass covered with earth at the root of some plant. During April and early in May, when the willows are in blossom, we have found the young recently hatched larvæ in considerable abundance creeping briskly over the bees, or with their heads plunged between the segments of the body, greedily sucking in the juices of their host. Those that we saw occurred on the Humble and other wild bees, and on various flies (Syrphus and Muscidæ), and there is no reason why they should not infest the Honey bee, which frequents similar flowers, as they are actually known to do in Europe. These larvæ are probably hatched out near where the bees hibernate, so as to creep into their bodies before they fly in the spring, as it would be impos-

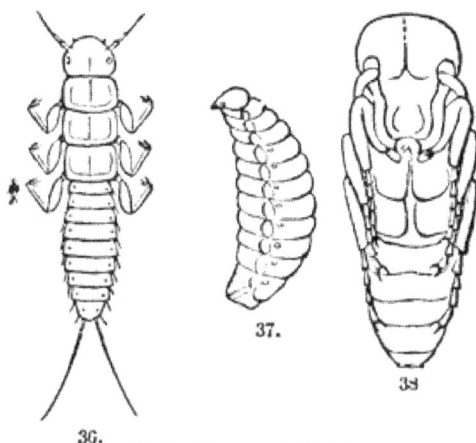

36.
37.
38.
Early Stages of Meloë.

sible for them to crawl up a willow tree ten feet high or more, their feet being solely adapted for climbing over the hairy body of the bee, which they do not leave until about to undergo their strange and unusual transformations.

In Europe, Assmuss states that on being brought into the nest by the bee, they leave the bee and devour the eggs in the bee cells, and then attack the bee bread. When full-fed and ready to pass through their transformations to attain the beetle state, instead of at once assuming the pupa and imago forms, as in the Trichodes represented in fig. 34, they pass through a *hyper-metamorphosis*, as Fabre, a French naturalist, calls it. In other words, the changes in form which are preparatory to assuming the pupa and imago states are more marked and almost coequal with the larva and pupa states, so that the Meloë, instead of passing through three states (the egg, larva and pupa), in reality passes through these and two others in addition, which are intermediate. The whole subject of the metamorphosis of this beetle needs revision, but Fabre states that the larva, soon after entering the nest of its host, changes its skin and assumes a second larva form. Newport, who with Siebold has carefully described the metamorphoses of Meloë, does not mention this stage in its development, which Fabre calls "pseudo-chrysalis." It is motionless, the head is mask-like, without movable appendages, and the feet are represented by six tubercles. This is more properly speaking the semi-pupa, and the mature pupa grows beneath its mask-like form, which is finally moulted. This form, however, according to Fabre, changes its skin and turns into a third larva form (Fig. 37). After some time it assumes its true pupa form (Fig. 38), and finally moults this skin to appear as a beetle.

Fabre has also, in a lively and well-written account, given a history of Sitaris, a European beetle, somewhat resembling Meloë. He states that Sitaris lays its eggs near the entrance of bees' nests, and at the very moment that the bee lays her egg in the honey cell, the flattened, ovate Sitaris larva drops from the body of the bee upon which it has been living, and feasts upon the contents of the freshly laid egg. After eating this delicate morsel it devours the honey in the cells of the bee and changes into a white, cylindrical, nearly footless grub, and after it is full-fed, and has assumed a supposed "pupa" state, the skin, without bursting, incloses a kind of hard "pupa" skin, which is very similar in outline to the former larva, within whose skin is found a whitish larva which directly changes into the true pupa. In a succeeding state this pupa in the ordinary way changes to a beetle which belongs to the same group of Coleoptera as Meloë. We cannot but think, from observations made on the humble bee, the wasp, two species of moths and several other insects, that this "hyper-metamorphosis" is not so abnormal a mode of insect metamorphosis as has been supposed, and that the changes of these insects, made beneath the skin of the mature larva before assuming the pupa state, are almost as remarkable as those of Meloë and Sitaris, though less easily observed than they. Several other beetles allied to Meloë are known to be parasitic on wild bees, though the accounts of them are fragmentary.

The history of Stylops, a beetle allied to Meloë, is no less strange than that of Meloë, and is in some respects still more interesting. On June 18th I captured an Andrena vicina which had been "stylopized." On looking at my capture I saw a pale reddish-brown triangular mark on the bee's abdomen; this was the flattened head and thorax of a female Stylops (Fig. 39a, position of the female of Stylops, seen in profile in the abdomen of the bee; Fig. 39b, the female seen from above. The head and thorax are soldered into a single flattened mass, the baggy hindbody being greatly enlarged like that of the gravid female of the white ant, and consisting of nine segments).

On carefully drawing out the whole body (Pl. 1, Fig. 6, as seen from above, and showing the alimentary canal ending in a blind sac; Fig. 6a, side view), which is very extensible, soft and baggy, and examining it under a high power of the microscope, we saw multitudes, at least several hundred, of very minute larvæ, like particles of dust to the naked eye, issuing in every direction from the body of the parent now torn open in places, though most of them made their exit through an opening on the under side of the head-thorax. The Stylops, being hatched while still in the body of the parent, is therefore viviparous. She probably never lays eggs.

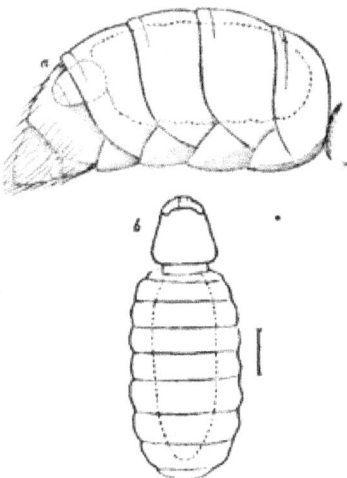

39. Female Stylops.

On the last of April, when the Mezereon was in blossom, I caught the singular looking male (Stylops Childreni, Fig. 40; a, side view; it is about one-fourth of an inch long), which was as unlike its partner as possible. I laid it under a tumbler, when the delicate insect flew and tumbled about till it died of exhaustion in a few hours.

It appears, then, that the larvæ are hatched during the middle or last of June from eggs fertilized in April. The larvæ then crawl out upon the body of the bee, on which they are transported to the nest, where they enter, according to Peck's observations, the body of the larva, on whose fatty parts they feed.

Previous to changing to a pupa the larva lives with its head turned towards that of its host, but before assuming the perfect state (which they do in the late summer or autumn) it must reverse its position. The female protrudes the front part of her body between the segments of the abdomen of her host, as represented in our figure. This change, Newport thinks, takes place after the bee-host has undergone its metamorphoses, though the bee does not leave her earthen cells until the following spring. Though the male Stylops deserts his host, his wingless partner is imprisoned during her whole life within her host, and dies immediately after giving birth to her myriad (for Newport thinks she produces over two thousand) offspring.

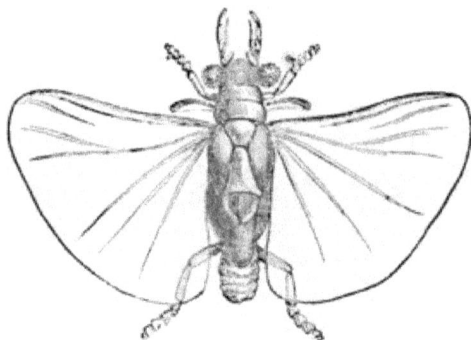

40. Male Stylops.

Xenos Peckii, an allied insect, was discovered by Dr. Peck to be parasitic in the body of wasps, and there are now known to be several species of this small but curious family, Stylopidæ, which are known to live parasitically on the bodies of our wild bees and wasps. The presence of these parasites finally exhausts the host, so that the sterile female bee dies prematurely.

As in the higher animals, bees are afflicted with parasitic worms which induce disease and sometimes death. The well-known hair worm, Gordius, is an insect parasite. The adult form is about the size of a slender knitting needle, and is seen in moist soil and in pools. It lays, according to Dr. Leidy, "millions of eggs connected together in long cords." The microscopical, tadpole-shaped young penetrate into the bodies of insects frequenting damp localities. Fairly ensconced within the body of their unsuspecting host, they luxuriate on its fatty tissues, and pass through their metamorphoses into the adult form, when they desert their living house and take to the water to lay their eggs. In Europe, Siebold has described Gordius subbifurcus, which infests the drones of the Honey bee, and also other insects. Professor Siebold has also described Mermis albicans, which is a similar kind of hair worm, from two to five inches long, and whitish in color. This worm is also found, strangely enough, only in the drones, though it is the workers which frequent watery places to appease their thirst.

Thousands of insects are carried off yearly by parasitic fungi. The ravages of the Muscardine, caused by a minute fungus (Botrytris Bassiana), have threat- ened the extinction of silk cul- ture in Europe, and the still more formidable disease called *pebrine* is thought to be of veg- etable origin. Dr. Leidy men- tions a fungus which must annu- ally carry off myriads of the Seventeen Year Locust. A some- what similar fungus, Mucor mel- litophorus (Fig. 41), infests bees, filling the stomach with microscopical colorless spores, so as greatly to weaken the in- sect.

As there is a probability that many insects, parasites on the

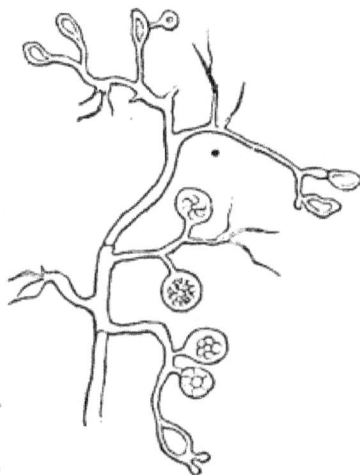

41. Bee fungus.

wild bees, may sooner or later afflict the Honey bee, and also to illustrate farther the complex nature of insect parasitism, we will for a moment look at some other bee parasites.

Among the numerous insects preying in some way upon the Humble bee are to be found other species of bees and moths, flies and beetles. Insect parasites often imitate their host: Apathus (Plate 1. Fig. 1, A. Ashtoni) can scarcely be distin- guished from its host, and yet it lives cuckoo-like in the cells of the Humble bee, though we know not yet how injurious it really is. Then there are Conops and Volucella, the former

Pl. 1.

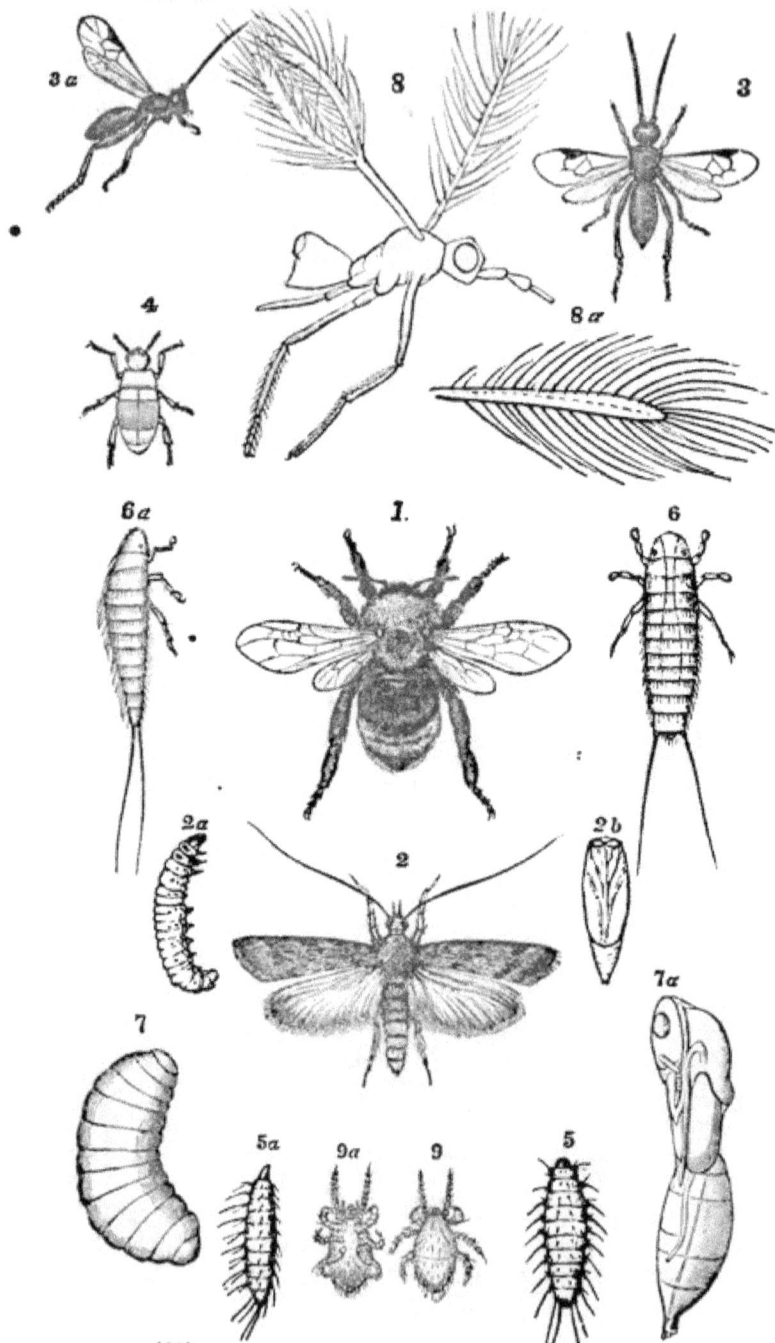

PARASITES OF BEES.

of which lives like Tachina and Phora within the bee's body,
while the latter devours the brood. The young (Plate 1, Figs.
5, 5 a) of another fly allied to Anthomyia, of which the Onion
fly (Fig. 42) is an example, is also not unfrequently met with.
A small beetle (Plate 1, Fig. 4, Antherophagus ochraceus) is a
common inmate of Humble bees' nests, and probably feeds upon
the wax and pollen. We have also found several larvæ (Fig.
43) of a beetle of which we do not know the adult form. Of
similar habits is probably a small moth (Nephopteryx Edmandsii,
Plate 1, Figs. 2; 2 a, larva; Fig. 2 b, chrysalis, or pupa) which
undoubtedly feeds upon the waxen walls of the bee cells, and
thus, like the attacks of the common bee moth (Galleria cere-
ana), whose habits are so well known as not to detain us, must
prove very prejudicial to the well being of the colony. This moth
is in turn infested by an Ichneumon fly (Microgaster nephopter-
icis, Plate 1, Figs. 3, 3 a) which must prove quite destructive.

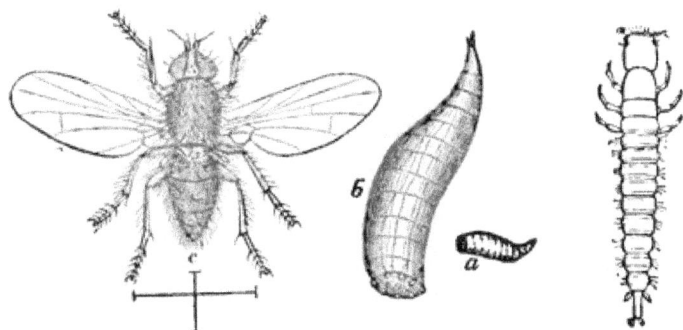

42. Onion Fly and Maggot.     43. Larva of Beetle.

The figures of the early stages of a minute ichneumon represented on the same plate (Fig. 7, larva, and 7 a, pupa, of Anthophorabia megachilis) which is parasitic on Megachile, the
Leaf-cutter bee, illustrates the transformations of the Ichneumon
flies, the smallest species of which yet known (and we believe
the smallest insect known at all) is the Pteratomus Putnami
(Pl. 1, Fig. 8, wanting the hind leg), or "winged atom," which
is only one-ninetieth of an inch in length, and is parasitic on
Anthophorabia, itself a parasite. A species of mite (Plate 1,
Figs. 9; 9 a, the same seen from beneath) is always to be found
in humble bees' nests, but it is not thought to be specially obnoxious to the bees themselves, though several species of mites
(Gamasus, etc.) are known to be parasitic on insects.

5

# CHAPTER IV.

## A FEW WORDS ABOUT MOTHS.

THE butterflies and moths from their beauty and grace, have always been the favorites among amateur entomologists, and rare and costly works have been published in which their forms and gorgeous colors are represented in the best style of natural history art. We need only mention the folio volume of Madam Merian of the last century, Harris's Aurelian, the works of Cramer, Stoll, Drury, Hübner, Horsfield, Doubleday and Westwood, and Hewitson, as comprising the most luxurious and costly entomological works.

Near the close of the last century, John Abbot went from London and spent several years in Georgia, rearing the larger and more showy butterflies and moths, and painting them in the larva, chrysalis and adult, or imago stage. These drawings he sent to London to be sold. Many of them were collected by Sir James Edward Smith, and published under the title of " The Natural History of the Rarer Lepidopterous Insects of Georgia, collected from the Observations of John Abbot, with the Plants on which they Feed." (London, 1797. 2 vols., fol.) Besides these two rare volumes there are sixteen folio volumes of drawings by Abbot in the Library of the British Museum. This work is of especial interest to the American student as it illustrates the early stages of many of our butterflies and moths.

Indeed the study of insects possesses most of its interest when we observe their habits and transformations. Caterpillars are always to be found, and with a little practice are easy to raise; we would therefore advise any one desirous of beginning the study of insects to take up the butterflies and moths. They are perhaps easier to study than any other group of insects, and are more ornamental in the cabinet. As a scientific study we would recommend it to ladies as next to botany in interest and in the ease in which specimens may be collected and examined. The example of Madam Merian, and several ladies in this country who have greatly aided science by their well filled cabinets, and critical knowledge of the various species and their transformations, is an earnest of what may be expected from their followers. Though the moths are easy to study compared with the bees, flies, beetles and bugs, and dragon flies, yet many questions of great interest in philosophical entomology have been answered by our knowledge of their structure and mode of growth. The great works of Herold on the evolution of a caterpillar; of Lyonet on the anatomy of the Cossus; of Newport on that of the Sphinx; and of Siebold on the parthenogenesis of insects, are proofs that the moths have engaged the attention of some of the master minds in science.

The study of the transformations of the moths is also of great importance to one who would acquaint himself with the questions concerning the growth and metamorphoses and origin of animals. We should remember that the very words "metamorphosis" and "transformation," now so generally applied to other groups of animals and used in philosophical botany, were first suggested by those who observed that the moth and butterfly attain their maturity only by passing through wonderful changes of form and modes of life.

The knowledge of the fact that all animals pass through some sort of a metamorphosis is very recent in physiology. Moreover the fact that these morphological eras in the life of an individual animal accord most unerringly with the gradation of forms in the type of which it is a member, was the discovery of the eminent physiologist Von Baer. Up to this time the true significance of the luxuriance and diversity of larval forms had never seriously engaged the attention of systematists in entomology.

What can possibly be the meaning of all this putting on and taking off of caterpillar habiliments, or in other words, the process of moulting, with the frequent changes in ornamentation, and the seeming fastidiousness and queer fancies and strange conceits of these young and giddy insects seems hidden and mysterious to human observation. Indeed, few care to spend the time and trouble necessary to observe the insect through its transformations; and that done, if only the larva of the perfect insect can be identified and its form sketched how much was gained! A truthful and circumstantial biography, in all its relations, of a single insect has yet to be written!

We should also apply our knowledge of the larval forms of insects to the details of their classification into families and genera, constantly collating our knowledge of the early stages with the structural relations that accompany them in the perfect state.

The simple form of the caterpillar seems to be a concentration of the characters of the perfect insect, and presents easy characters by which to distinguish the minor groups; and the relative rank of the higher divisions will only be definitely settled when their forms and methods of transformation are thoroughly known. Thus, for example, in two groups of the large Attacus-like moths, which are so amply illustrated in Dr. Harris's "Treatise on Insects injurious to Vegetation;" if we take the different forms of the caterpillars of the Tau moth of Europe, which are figured by Duponchel and Godard, we find that the very young larva has four horn-like processes on the front, and four on the back part of the body. The full grown larva of the Regalis moth, of the Southern and Middle states, is very similarly ornamented. It is an embryonic form, and therefore inferior in rank to the Tau moth. Multiply these horns over the surface of the body, lessen their size, and crown them with hairs, and we have our Io moth, so destructive to corn. Now take off the hairs, elongating and thinning out the tubercles, and make up the loss by the increased size of the worm, and we have the caterpillar of our common Cecropia moth. Again, remove the naked tubercles almost wholly, smooth off the surface of the body, and contract its length, thus giving a greater convexity and angularity to the rings, and we have before us the larva of the stately Luna moth that tops this royal family. Here are certain criteria for placing these insects before our minds in the order that nature has placed them. We have certain facts for determining which of these three insects is highest and which lowest in the scale, when we see the larva of the Luna moth throwing off successively the Io and Cecropia forms to take on its own higher features. So that there is a meaning in all this shifting of insect toggery.

This is but an example of the many ways in which both pleasure and mental profit may be realized from the thoughtful study of caterpillar life.

In collecting butterflies and moths for cabinet specimens, one needs a gauze net a foot and a half deep, with the wire frame a foot in diameter; a wide-mouthed bottle containing a parcel of cyanide of potassium gummed on the side, in which to kill the moths, which should, as soon as life is extinct, be pinned in a cork-lined collecting box carried in the coat pocket. The captures should then be spread and dried on a grooved setting board, and a cabinet formed of cork-lined boxes or drawers; as a substitute for cork, frames with paper tightly stretched over them may be used, or the pith of corn-stalks or palm wood. Caterpillars should be preserved in spirits, or in glycerine with a little alcohol a d d e d.

Some persons ingeniously empty the skins and inflate them over a flame so that they may be pinned by the side of the adult.

Some of the most troublesome and noxious in-, sects are found among the moths. I need only mention the canker worm and American tent caterpillar, and the various kinds of cut worms, as instances.

We must not, however, forget the good done by insects. They undoubt-

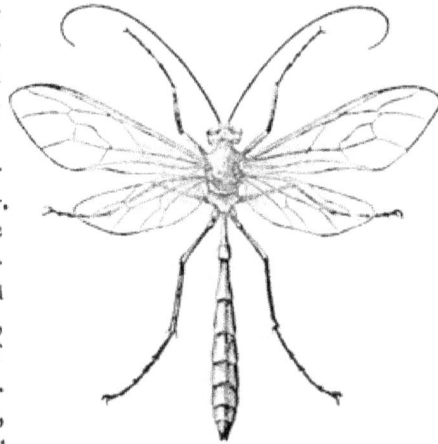

43. Parasite of the American Silk Worm.

edly tend by their attacks to prevent an undue growth of vegetation. The pruning done to a tree or herb by certain insects undoubtedly causes a more healthy growth of the branches and leaves, and ultimately a greater production of fruit. Again, as pollen-bearers, insects are a most powerful agency in nature. It is undoubtedly the fact that the presence of bees in orchards increases the fruit crop, and thus the thousands of moths (though injurious as caterpillars), wild bees and other insects, that seem to live without purpose, are really, though few realize it, among the best friends and allies of man.

Moreover, insects are of great use as scavengers; such are the young or maggots of the house fly, the mosquitoes, and numerous other forms, that seem created only to vex us when in the winged state.    Still a larger proportion of insects are directly beneficial from their habit of attacking injurious species, such as the ichneumons (Fig. 43, the ichneumon of the American silk worm) and certain flies (Fig. 44, Tachina); also many carnivorous species of w a s p s beetles and flies, d r a g o n flies and Aphis lions (Fig. 45, the l a c e - w i n g e d fly; adult, larva and eggs).

But few, however, suspect how enormous are the losses to crops in this coun try entailed by the attacks of the injurious  s p e c i e s.

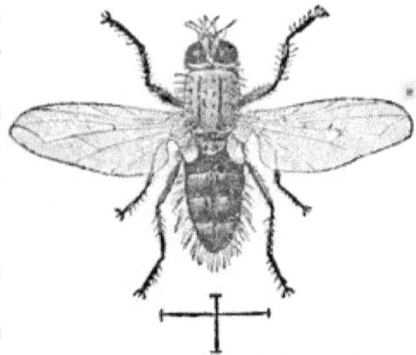

44. Tachina, parasite of Colorado Potato Beetle.

In Europe, the subject of applied entomology has always attracted a great deal of attention.    Most sumptuons works, elegant quartos prepared by naturalists known the world over, and published at government expense, together with smaller treatises, have frequently appeared; while the subject

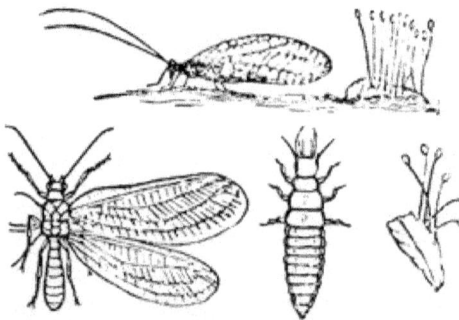

45. The Lace-winged Fly, its Larva and Eggs.

is taught in the numerous agricultural colleges and schools, especially of Germany.

In the densely populated countries of Europe, the losses occasioned by injurious insects are most severely felt, t h o u g h from  m a n y  causes, such as the greater abundance of their insect parasites, and the far greater care taken by the people to exterminate their insect enemies, they have not proved so destructive as in our own land.

In this connection I may quote from one of Dr. Asa Fitch's reports on the noxious insects of New York, where he says : "I find that in our wheat-fields here, the midge formed 59 per cent.

of all the insects on this grain the past summer; whilst in France, the preceding summer, only 7 per cent. of the insects on wheat were of this species. In France the parasitic destroyers amounted to 85 per cent.; while in this country our parasites form only 10 per cent."

A true knowledge of practical entomology may well be said to be in its infancy in our own country, when, as is well known to agriculturists, the cultivation of wheat has almost been given up in New England, New York, Pennsylvania, Ohio and Virginia, from the attacks of the wheat midge, Hessian fly, joint worm, and chinch bug. According to Dr. Shimer's estimate, says Mr. Riley, in his Second Annual Report on the Injurious Insects of Missouri, which may be considered a reasonable one, "in the year 1864 three-fourths of the wheat, and one-half of the corn crop were destroyed by the chinch bug throughout many extensive districts, comprising almost the entire North-West. At the annual rate of increase, according to the United States Census, in the State of Illinois, the wheat crop ought to have been about thirty millions of bushels, and the corn crop about one hundred and thirty-eight million bushels. Putting the cash value of wheat at $1.25, and that of corn at 50 cents, the cash value of the corn and wheat destroyed by this insignificant little bug, no bigger than a grain of rice, in one single State and one single year, will therefore, according to the above figures, foot up to the astounding total of *over seventy-three millions of dollars!*"

The imported cabbage butterfly (Pieris rapæ), recently introduced from Europe, is estimated by the Abbé Provancher, a Canadian entomologist, to destroy annually two hundred and forty thousand dollars' worth of cabbages around Quebec. The Hessian fly, according to Dr. Fitch, destroyed fifteen million dollars' worth of wheat in New York State in one year (1854). The army worm of the North (Leucania unipuncta), which was so abundant in 1861, from New England to Kansas, was reported to have done damage that year in Eastern Massachusetts exceeding half a million of dollars. The joint worm (Isosoma hordei) alone sometimes cuts off whole fields of grain in Virginia and northward. The Colorado potato beetle is steadily moving eastward, now ravaging the fields in Indiana and Ohio, and only the forethought and ingenuity in devising means of checking its attacks, resulting from a thorough study of its habits, will deliver our wasted fields from its direful assaults.

These are the injuries done by the more abundant kinds of insects injurious to crops. We should not forget that each fruit or shade tree, garden shrub or vegetable, has a host of insects peculiar to it, and which, year after year, renew their attacks. I could enumerate upwards of fifty species of insects which prey upon cereals and grass, and as many which infest our field crops. Some thirty well known species ravage our garden vegetables. There are nearly fifty species which attack the grape vine, and their number is rapidly increasing. About seventy-five species make their annual onset upon the apple tree, and nearly an equal number may be found upon the plum, pear, peach and cherry. Among our shade trees, over fifty species infest the oak; twenty-five the elm; seventy-five the walnut, and over one hundred species of insects prey upon the pine.

Indeed, we may reasonably calculate the annual loss in our country alone, from noxious animals and the lower forms of plants, such as rust, smut and mildew, as (at a low estimate) not far from five hundred million dollars annually. Of this amount, at least one-tenth, or fifty million dollars, could probably be saved by human exertions.

To save a portion of this annual loss of food stuffs, fruits and lumber, should be the first object of farmers and gardeners. When this saving is made, farming will become a profitable and safe profession. But while a few are well informed as to the losses sustained by injurious insects, and use means to ward off their attacks, their efforts are constantly foiled by the negligence of their neighbors. As illustrated so well by the history of the incursions of the army worm and canker worm, it is only by a combination between farmers and orchardists that these and other pests can be kept under. The matter can be best reached by legislation. We have fish and game laws; why should we not have an insect law? Why should we not frame a law providing that farmers, and all owning a garden or orchard, should coöperate in taking preventive measures against injurious insects, such as early or late planting of cereals, to avert the attacks of the wheat midge and Hessian fly; the burning of stubble in the autumn and spring to destroy the joint worm; the combined use of proper remedies against the canker worm, the various cut worms, and other noxious caterpillars? A law carried out by a proper State entomological constabulary, if it may be so designated, would compel the idle and shiftless to clear their farms and gardens of noxious animals.

Among some of the injurious insects reported on by Mr. Riley, the State Entomologist of Missouri, is a new pest to the cucumber in the West, the Pickle worm (Phacellura nitidalis, Fig. 46).

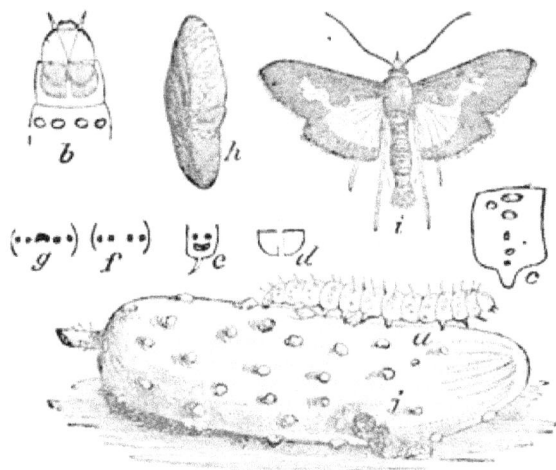

46. Pickle Worm and its Moth.

This is a caterpillar which bores into the cucumbers when large enough to pickle, and which is occasionally found in pickles. Three or four worms sometimes occur in a cucumber, and in the garden a single one will cause it to rot. One of the most troublesome intruders in our graperies is the Vine dresser (Chœrocampa pampinatrix, Fig. 47, larva and pupa; Fig. 48, adult), a single caterpillar of which will sometimes "strip a small vine of its leaves in a few nights," and occasionally nips off bunches of half-grown grapes.

Another caterpillar, which is sometimes so abundant as nearly to defoliate the grape vine, is the eight spotted Alypia (Fig. 49; a, larva; b, side view of a segment). This must not be confounded with the bluish larva of the Wood Nymph, Eudryas grata (Fig. 50), which differs from the Alypia caterpillar in being bluish, and in wanting the white patches on the side of the body, and the more prominent hump on the end of the body.

48. Vine Dresser Moth.

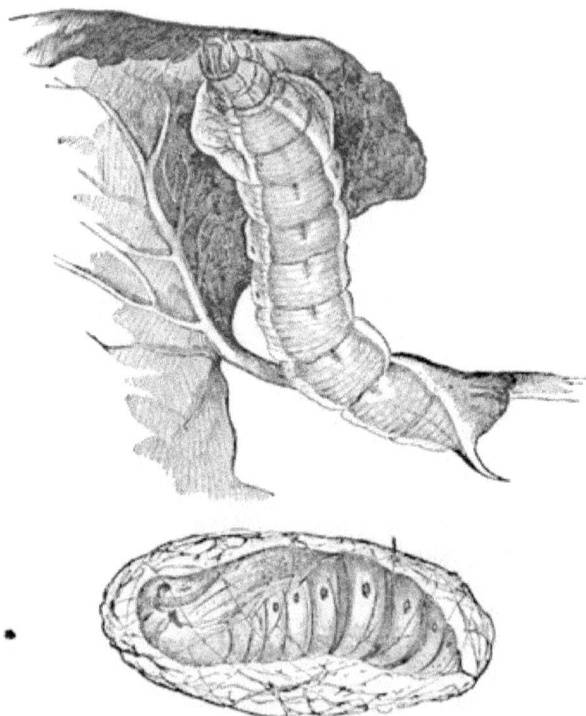

47. Vine Dresser and Chrysalis.

GRAPE MOTHS.

59

Another moth (Psychomorpha epimenis, Fig. 51, *a*, larva; *b*, side view of a segment; *c*, top view of the hump), also feeds on the grape, eating the terminal buds. It is also bluish, and wants the orange bands on the side of the body. Another moth of this family is the American Procris (Acoloithus Americana, Fig. 52*a*, larva; *b*, pupa; *c*, cocoon; *d*, *e*, imago); a dark blue moth, with a deep orange collar, whose black and yellow caterpillar is gregarious (Fig. 53), living in companies of a dozen or more and eating the

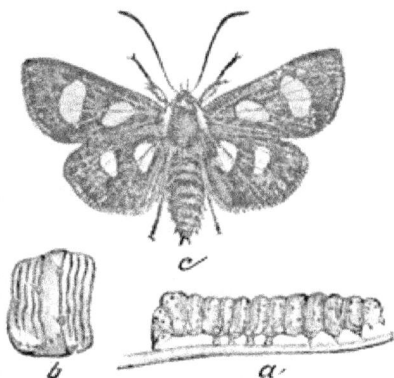

49. Eight-spotted Alypia and Larva.

softer parts of the leaves. It is quite common in the Western and Southern States. The figure represents two separate broods of caterpillars feeding on either side of the midrib of the leaf.

50. Eudryas grata.

But if the moths are, as a rule, the enemies of our crops, there are the silk worms of the East and Southern Europe and California, which afford the means of support to multitudes of the poorer classes, and supply one of the most valuable articles of clothing. Blot out the silk worm, and we should

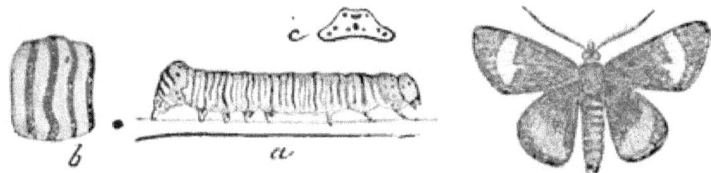

51. Larva of Psychomorpha.

remove one of the most important sources of national wealth, the annual revenue from the silk trade of the world amounting to $254,500,000.

Silk culture is rapidly assuming importance in California, and though the Chinese silk worm has not been successfully cultivated in the Eastern States. yet the American silk worm, Teleas Polyphemus (see frontispiece, male; Fig. 54, larva; 55, pupa; 56, cocoon), can, we are assured by Mr. Trouvelot, be made a source of profit.

This is a splendid member of the group of which the gigantic Attacus Atlas of China is a type. It is a large, fawn colored moth with a tawny tinge; the caterpillar is pale green, and is of the size indicated

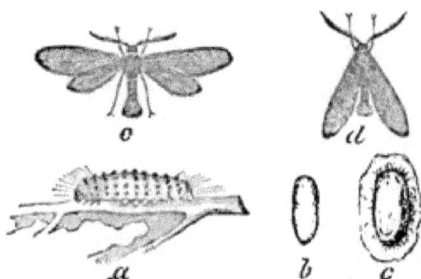

52. American Procris and Young.

in the cut. Mr. Trouvelot says that of the several kinds of silk worms, the larva of the present species alone deserves attention. The cocoons of Platysamia Cecropia may be rendered of some commercial value, as the silk can be carded, but the chief objection is the difficulty of raising the larva.

"The Polyphemus worm spins a strong, dense, oval cocoon, which is closed at each end, while the silk has a very strong and glossy fibre." Mr. Trouvelot, from whose

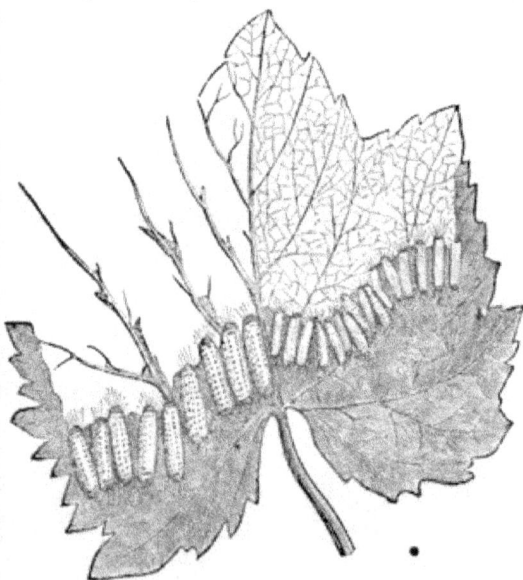

53. Larvæ of American Procris.

interesting account in the first volume of the "American Naturalist" we quote, says that in 1865 " not less than a million could be seen feeding in the open air upon bushes covered with a net; five acres of woodland were swarming with caterpillar life."

The bushes were scrub oaks, the worms being protected by a net. After meeting with such great success Mr. Trouvelot lost all his worms by pebrine, the germs being imported in eggs received from Japan through M. Guérin-Méneville of Paris. Enough, however, was done to prove that silk raising can be carried on profitably, when due precautions are taken, as far north as Boston. As this moth extends to the tropics, it can be reared with greater facility southwards. The cocoon is strong and dense, and closed at each end, so that the thread is c o n t i n u o u s, while the silk has a very strong and glossy fibre.

Next in value to the American silk worm, is the Ailanthus silk worm (Samia Cynthia) a species allied to our Callosamia P r o m e t h e a. It originated from China, where it is cultivated, and was introduced into Italy in 1858, and thence spread into France, where it was i n t r o d u c e d by M. Guérin-Méneville. Its silk is said to be much stronger than the fibre of cotton, and is a mean between fine wool and ordinary silk. The worm is very hardy, and can be reared in the open air both in this country and in Europe. The main drawback to its culture is the difficulty in unreeling the tough cocoon, and the shortness of the thread, the cocoon being open at one end.

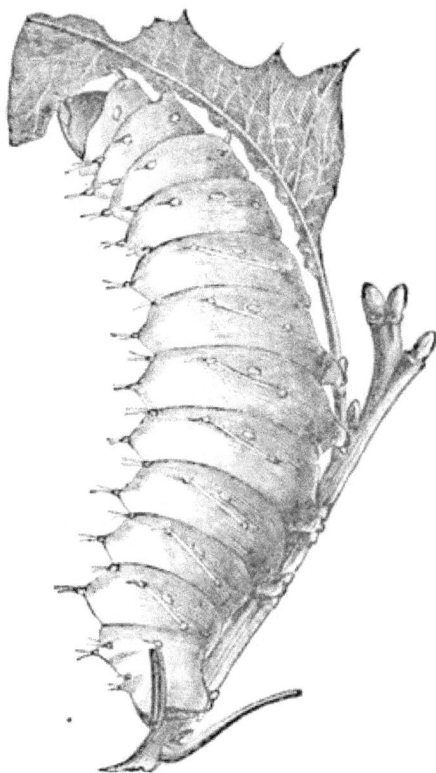

54. American Silk Worm.

The Yama-maï moth (Antheræa Yama-maï) was introduced into France from Japan in 1861. It is closely allied to the Polyphemus moth, and its caterpillar also feeds on the oak. Its

6

silk is said to be quite brilliant, but a little coarser and not so strong as that of the Bombyx mori. The Perny silk worm is extensively cultivated by the Chinese in Manchouria, where it

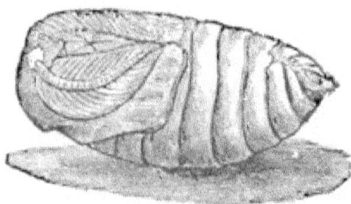

feeds on the oak. Its silk is coarser than that of the common silk worm, but is yet fine, strong and glossy. Bengal has furnished the Tussah moth, which lives in India on the oak and a variety of other trees. It is largely raised in French and English India, according to

55. Chrysalis of American Silk Worm.

Noguès, and is used in the manufacture of stuffs called corahs.

The last kind of importance is the Arrhindy silk worm, from India. It has been naturalized in France and Algeria by M. Guérin-Méneville, who has done so much in the application of entomology to practical life. It is closely allied to the Cynthia or Ailanthus worm, with the same kind of silk and a similar cocoon, and feeds on the castor oil plant.

The diseases of silk worms naturally receive much attention. Like those afflicting mankind, they arise from bad air, resulting from too close confinement, bad food, and other adverse causes. The most fatal and wide-spread disease, and one which since 1854 has threatened the extermination of silk worms in Europe, is the *pebrine*. It is due to the presence of minute vegetable corpuscles, which attack both the worms and the eggs. It was this disease which swept off thousands of Mr. Trouvelot's Polyphemus worms, and put a sudden termination to his important experiments, the germs having been implanted in eggs of the Yamamai moth imported from Japan by M. Guérin-Méneville,

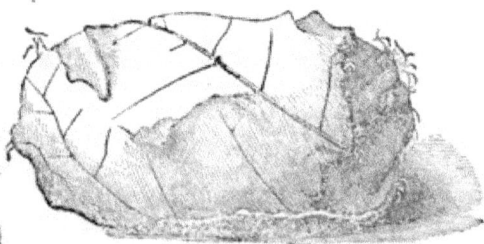

56. Cocoon of American Silk Worm.

and which were probably infected as they passed through Paris. Though the disaster happened several years since, he tells us that it will be useless for him to attempt the raising of silk worms in the town where his establishment is situated, as the germs of the disease are most difficult to eradicate.

So direful in France were the ravages of this disease that two of the most advanced naturalists in France, Quatrefages and Pasteur, were commissioned by the French government to investigate the disease. Pasteur found that the infected eggs differed in appearance from the sound ones, and could thus be sorted out by aid of the microscope and destroyed. Thus these investigations, carried on year after year, and seeming to the ignorant to tend to no practical end, resulted in saving to France her silk culture. During the past year (1871) so successful has his method proved that a French scientific journal expresses the hope of the complete reëstablishment and prosperity of this great industry. A single person who obtained in 1871 in his nurseries 50,000 ounces of eggs, hopes the next year to obtain 100,000 ounces, from which he expects to realize about one million dollars.

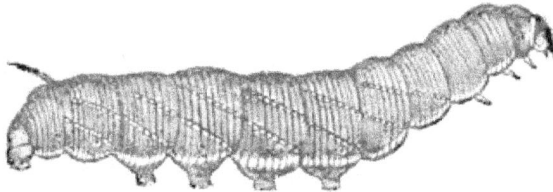

The Potato Caterpillar.

# CHAPTER V.

## THE CLOTHES MOTH.

For over a fortnight we once enjoyed the company of the caterpillar of a common clothes moth. It is a little pale, delicate worm (Fig. 57, magnified), about the size of a darning needle, and rather less than half an inch in length, with a pale horn-colored head, the ring next the head being of the same color. It has sixteen feet, the first six of them well developed and constantly in use to draw the slender body in and out of its case. Its head is armed with a formidable pair of jaws, with which, like a scythe, it mows its way through thick and thin.

But the case is the most remarkable feature in the history of this caterpillar. Hardly has the helpless, tiny worm broken out of the egg, previously laid in some old garment of fur or wool, or perhaps in the haircloth of a sofa, when it begins to make a shelter by cutting the woolly fibres or soft hairs into bits, which it places at each end in successive layers, and, joining them together by silken threads, constructs a cylindrical tube (Fig. 58) of thick, warm felt, lined within with the finest silk the tiny worm can spin. The case is not perfectly cylindrical, being flattened slightly in the middle, and contracted a little just before each end, both of which are always kept open. The case before us is of a stone-gray color, with a black stripe along the middle, and with rings of the same color round each opening. Had the caterpillar fed on blue or yellow cloth, the case would, of course, have been of those colors. Other cases, made by larvæ which had been eating loose cotton, were quite irregular in form, and covered loosely with bits of cotton thread, which the little tailor had not trimmed off.

Days go by. A vigorous course of dieting on its feast of

wool has given stature to our hero. His case has grown uncom-
fortably small. Shall he leave it and make another? No housewife is more prudent and saving. Out come those scissor-jaws,
and, lo! a fearful rent along each side of one end of the case.
Two wedge-shaped patches mend the breach; the caterpillar
retires for a moment and reappears at the other end; the scis-
sors are once more pulled out; two rents appear, to be filled up
by two more patches or gores, and our caterpillar once again
breathes more freely, laughs and grows fat upon horse hair and
lambs' wool. In this way he enlarges his case till he stops
growing.

Our caterpillar seeming to be full-grown, and apparently out
of employment, we cut the end of his case half off. Two or
three days after, he had mended it from the inside, drawing the
two edges together by silken threads, and, though he had not
touched the outside, yet so
neatly were the two parts
joined together that we had
to search for some time,
with a lens, to find the scar.

To keep our friend busy
during the cold, cheerless
weather, for it was mid-
winter, we next cut a third
of the case entirely off. Nothing daunted, the little fel-
low bustled about, drew in a mass of the woolly fibres, filling
up the whole mouth of his den, and began to build on afresh,
and from the inside, so that the new-made portion was smaller
than the rest of the case. The creature worked very slowly,
and the addition was left in a rough, unfinished state.

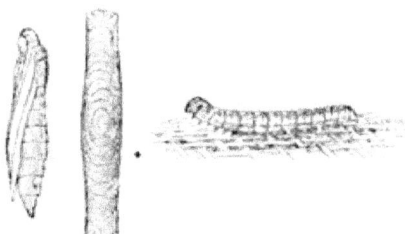

59.  58.  57.

Early stages of the Clothes Moth.

We could easily spare these voracious little worms hairs
enough to serve as food, and to afford material for the construc-
tion of their paltry cases; but that restless spirit that ever
urges on all beings endowed with life and the power of motion,
never forsakes the young clothes moth for a moment. He
will not be forced to drag his heavy case over rough hairs and
furzy wool, hence with his keen jaws he cuts his way through.
Thus, the more he travels, the more mischief he does.

After taking his fill of this sort of life he changes to a chrysalid (Fig. 59), and soon appears as one of those delicate, tiny,

demure moths that fly in such numbers from early in the spring until the autumn.

Very many do not recognize these moths in their perfect stage, so small are they, and vent their wrath on those great millers that fly around lamps in warm summer evenings. It need scarcely be said that these large millers are utterly guiltless of any attempts upon our wardrobes; they make their attacks in a more open form on our gardens and orchards.

We will give a more careful description of the clothes moth, which was found in its different stages June 12th in a mass of loose cotton. The larva is white, with a tolerably plump body, which tapers slightly towards the tail, while the head is much of the color of gum-copal. The rings of the body are thickened above, especially on the thoracic ones, by two transverse thickened folds. It is one-fifth of an inch long.

The body of the chrysalis, or pupa, is considerably curved, with the head smooth and rounded. The long antennæ, together with the hind legs, which are folded along the breast, reach to the tip of the hind body, on the upper surface of each ring of which is a short transverse row of minute spines, which aid the chrysalis in moving towards the mouth of its case, just before the moth appears. At first the chrysalis is whitish, but just before the exclusion of the moth becomes the color of varnish.

When about to cast its pupa skin, the skin splits open on the back, and the perfect insect glides out. The act is so quickly over with, that the observer has to look sharp to observe the different steps in the operation.

Our common clothes moth (Tinea flavifrontella. Fig. 60) is of a uniform light-buff color, with a silky iridescent lustre, the hind wings and abdomen being a little paler. The head is

60. Clothes Moth.

thickly tufted with hairs and is a little tawny, and the upper side of the densely hirsute feelers (palpi) is dusky. The wings are long and narrow, with the most beautiful and delicate long silken fringe, which increases in length towards the base of the wing.

They begin to fly in May, and last all through the season, fluttering with a noiseless, stealthy flight in our apartments, and laying their eggs in our woollens.

Successive broods of the clothes moth appear through the summer. In the autumn they cease eating, retire within their cases, and early in spring assume the chrysalis state.

There are several allied species which have much the same habits, except that they do not all construct cases, but eat carpets, clothing, articles of food, grain, etc., and objects of natural history.

Careful housewives are not much afflicted with these pests. The slovenly and thriftless are overrun with them. Early in June woollens and furs should be carefully dusted, shaken and beaten. Dr. T. W. Harris states that "powdered black pepper, strewed under the edge of carpets, is said to repel moths. Sheets of paper sprinkled with spirits of turpentine, camphor in coarse powder, leaves of tobacco, or shavings of Russia leather, should be placed among the clothes when they are laid aside for the summer; and furs and other small articles can be kept by being sewed in bags with bits of camphor wood, red cedar, or of Spanish cedar; while the cloth lining of carriages can be secured forever from the attacks of moths by being washed or sponged on both sides with a solution of the corrosive sublimate of mercury in alcohol, made just strong enough not to leave a white stain on a black feather." The moths can be most readily killed by pouring benzine among them, though its use must be much restricted from the disagreeable odor which remains. The recent experiments made with carbolic acid, however, convince us that this will soon take the place of other substances as a preventive and destroyer of noxious insects.

The Juniper Sickle-wing.

# CHAPTER VI.

## THE MOSQUITO AND ITS FRIENDS.

THE subject of flies becomes of vast moment to a Pharaoh, whose ears are dinned with the buzz of myriad winged plagues, mingled with angry cries from malcontent and fly-pestered subjects; or to the summer traveller in northern lands, where they oppose a stronger barrier to his explorations than the loftiest mountains or the broadest streams; or to the African pioneer, whose cattle, his main dependence, are stung to death by the Tsetze fly; or the farmer whose eyes on the evening of a warm spring day, after a placid contemplation of his growing acres of wheat blades, suddenly detects in dismay clouds of the Wheat midge and Hessian fly hovering over their swaying tops. The subject, indeed, has in such cases a national importance, and a few words regarding the main points in the habits of flies — how they grow, how they do not grow (after assuming the winged state), and how they bite; for who has not endured the smart and sting of these dipterous Shylocks, that almost torment us out of our existence while taking their drop of our heart's blood — may be welcome to our readers.

The Mosquito will be our first choice. As she leaps off from her light bark, the cast chrysalis skin of her early life beneath the waters, and sails away in the sunlight, her velvety wings fringed with silken hairs, and her neatly bodiced trim figure (though her nose is rather salient, considering that it is half as long as her entire body), present a beauty and grace of form and movement quite unsurpassed by her dipterous allies. She draws near and softly alights upon the hand of the charmed beholder, subdues her trumpeting notes, folds her wings noiselessly upon her back, daintily sets down one foot after the other, and with

(68)

an eagerness chastened by the most refined delicacy for the
feelings of her victim, and with the air of Velpeau redivivus,
drives through crushed and bleeding capillaries, shrinking nerves
and injured tissues, a many-bladed lancet of marvellous fine-
ness, of wonderful complexity and fitness. While engorging
herself with our blood, we will examine under the microscope
the mosquito's mouth. The head (Fig. 61) is rounded, with
the two eyes occupying a large part of the surface, and nearly
meeting on the top of the head. Out of the forehead, so to
speak, grow the long, delicate, hairy antennæ ($a$), and just below arises the long beak which consists of the bristle-like maxillæ ($mx$, with their palpi, $mp$) and mandibles ($m$), and the
single hair-like labrum,
these five bristle-like organs being laid in the hollowed labium ($l$). Thus
massed into a single awl-
like beak, the mosquito,
without any apparent effort,
thrusts them all except the
labium into the flesh. Her
hind body may be seen filling with the red blood, until it cries quits, and the
insect withdraws its sting
and flies sluggishly away.
In a moment the wounded
parts itch slightly, though
a very robust person may

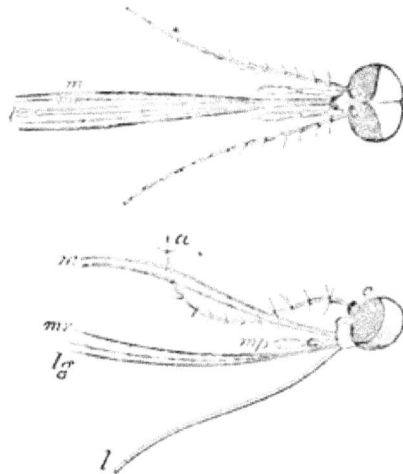

61. Head of the Mosquito.

not notice the irritation, or a more delicate individual if asleep;
though if weakened by disease, or if stung in a highly vascular
and sensitive part, such as the eyelid, the bite becomes really
a serious matter. Multiply the mosquito a thousand fold, and
one flees their attacks and avoids their haunts as he would a
nest of hornets. Early in spring the larva (Fig. 62, A) of the
mosquito may be found in pools and ditches. It remains at
the bottom feeding upon decaying matter (thus acting as a
scavenger, and in this state doing great benefit in clearing
swamps of miasms), until it rises to the surface for air, which
it inhales through a single respiratory tube ($c$) situated near
the tail. When about to transform into the pupa state, it

contracts and enlarges anteriorly near the middle, the larval skin is thrown off, and the insect appears in quite a different form (Fig. 62, B). The head and thorax are massed together, the rudiments of the mouth parts and of the wings and legs being folded upon the breast, while there are two breathing

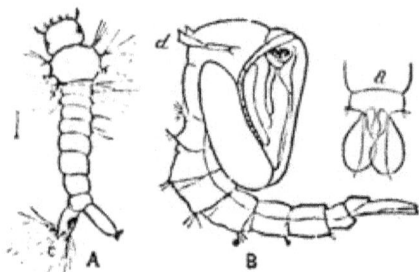

62. Larva and Pupa of the Mosquito.

tubes (d) situated upon the back instead of the tail, which ends in two broad paddles (a); so that it comes to the surface, head foremost instead of tail first, a position according better with its increased age and experience in pond life. In a few days the pupa skin is cast; the insect, availing itself of its old habiliments as a raft upon which to float while its body is drying, grows lighter, and its wings expand for its marriage flight. The males are beautiful, both physically and morally, as they do not bite; their manners are more retiring than those of their stronger minded partners, as they rarely enter our dwellings, and live unnoticed in the woods. They may be easily distinguished from the females by their long maxillary palpi, and their thick, bushy, feathered antennæ. The female lays her elongated, oval eggs in a boat-shaped mass, which floats on the water. A mosquito lives three or four weeks in the water before changing to the adult or winged stage. How many days they live in the latter state we do not know.

Our readers will understand, then, that all flies, like our mosquito for example, grow while in the larva and pupa state, *and after they acquire wings do not grow*, so that the small midges are not young mosquitoes, but the adult winged forms of an entirely different species and genus of fly; and the myriads of small flies, commonly supposed to be the young of larger flies, are adult forms belonging to different species of different genera, and perhaps of different families of the suborder of Diptera. The typical species of the genus Culex, to which the mosquito belongs, is Culex pipiens, described by Linnæus, and there are already over thirty North American species of this genus described in various works. Few insects live in the sea, but along the coast of New England a small, slender white larva (Fig. 63a, magnified, and head greatly enlarged; Fig. 64, pupa and fore foot of larva, showing the hooks), whose body is no thicker than a knitting needle, lives between tides, and has even been dredged at a depth of over a hundred feet, which transforms into a yellow mosquito-like fly (Fig. 65, with head of the female, magnified) which swarms in summer in immense numbers. I have called it provisionally Chironomus oceanicus, or Ocean gnat. The larvæ of other species have been found by Mr. S. I. Smith living at

65. Ocean Gnat.

great depths in our Northern lakes. These kinds of gnats are usually seen early in spring hovering in swarms in mid air.

The strange fact has been discovered by Grimm, a Russian naturalist, that the pupa of a feathered gnat is capable of laying eggs which produce young during the summer time. Previous to this it had been discovered that a larva of a gnat (Fig. 66, a, eggs from which the young are produced) which lives under the bark of trees in Europe, also produced young born alive.

The Hessian fly (Fig. 67, a, larva;

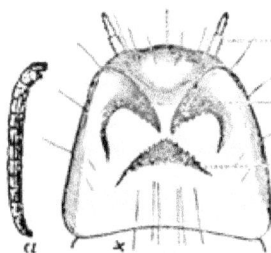

63. Larva of Ocean Gnat.

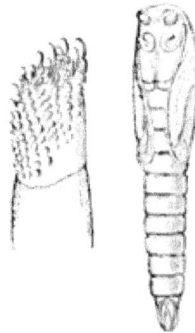

64. Pupa of Ocean Gnat.

b pupa; c, stalk of wheat injured by larvæ) and Wheat midge, which are allied to the mosquito, are briefly referred to in the calendar, so that we pass over these to consider another pest of our forests and prairies.

The Black fly is even a more formidable pest than the mosquito. In the northern, subarctic regions, it opposes a barrier against travel. The Labrador fisherman spends his summer on the sea shore, scarcely daring to penetrate the interior on account of the swarms of these flies. During a summer residence on this coast, we sailed up the Esquimaux river for six or eight miles, spending a few hours at a house situated on the bank. The day was warm and but little wind blowing, and the swarms of black flies were absolutely terrific. In vain we frantically waved our net among them, allured by some rare moth; after making a few desperate charges in the face of the thronging pests, we had to retire to the house, where the windows actually swarmed with them; but here they would fly in our faces, crawl under one's clothes, where they even remain and bite in the night. The children in the house were sickly and worn by their unceasing torments; and the shaggy Newfoundland dogs whose thick coats would seem to be proof against their bites ran from their shelter beneath the bench and dashed into the river, their only retreat. In cloudy weather, unlike the mosquito, the black fly disappears, only flying when the sun shines. The bite of the black fly is often severe, the creature leaving a large clot of blood to mark the scene of its surgical triumphs. Prof. E. T. Cox, State Geologist of Indiana, has sent us specimens of a much larger fly, which Baron Osten Sacken refers to this genus, which is called on the prairies, where it is said to bite horses to death, the Buffalo Gnat. Westwood states that an allied fly

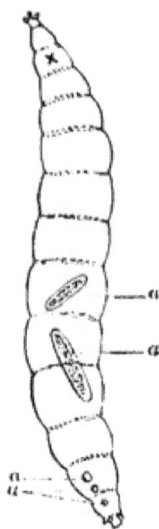

66. Viviparous gall larva.

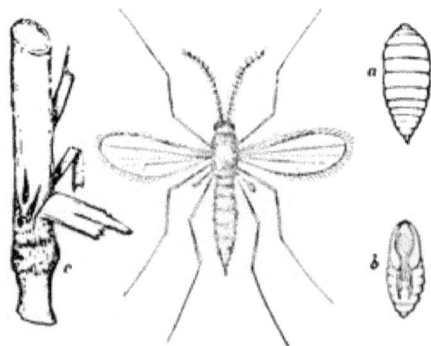

67. Hessian Fly and its Young.

(Rhagio Columbaschensis) is one of the greatest scourges of man and beast in Hungary, where it has been known to kill cattle.

The Simulium molestum (Fig. 68, enlarged), as the black fly is called, lives during the larva state in the water. The larva of a Labrador species (Fig. 69, enlarged) which we found, is about a quarter of an inch long, and of the appearance here indicated. The pupa is also aquatic, having long respiratory filaments attached to each side of the front of the thorax. According to Westwood, "the posterior part of its body is enclosed in a semioval membranous cocoon, which is at first formed by the larva, the anterior part of which is eaten away before changing to a pupa, so as to be open in front. The

68. Black Fly.

imago is produced beneath the surface of the water, its fine silky covering serving to repel the action of the water."

Multitudes of a long, slender, white worm may often be found living in the dirt, and sour sap running from wounds in the elm tree. Two summers ago we discovered some of these larvæ, and on rearing them found that they were a species of Mycetobia (Fig. 70; *a*, larva; *b*, pupa). The larva is remarkable for having the abdominal segments divided into two portions, the hinder much

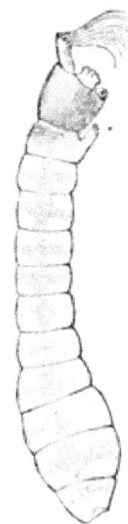

69. Black Fly Larva.

smaller than the anterior division. Its whole length is a little over a third of an inch. The pupæ were found sticking out in considerable numbers from the tree, being anchored by the little spines at the tail. The head is

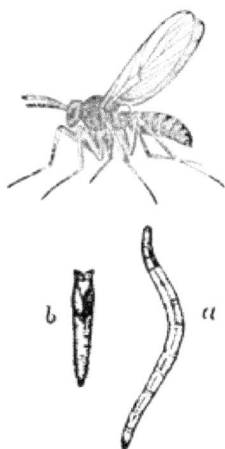

70. Mycetobia.

square, ending in two horns, and the body is straight and covered with spines, especially towards the end of the tail. They were a fifth of an inch in length. The last of June the flies appeared, somewhat resembling gnats, and about a line long. The worms continued to infest the tree for six weeks, the flies remaining either upon or near it.

7

We now come to that terror of our equine friends, the Horse
fly, Gad, or Breeze fly. In its larval state, some species live in
water, and in damp places under stones and pieces of wood,
and others in the earth away from water, where they feed on
animal, and, probably, on decaying matter. Mr. B. D. Walsh
found an aquatic larva of this genus, which, within a short time,
devoured eleven water snails. Thus at this stage of existence,
this fly, often so destructive, even at times killing our horses, is
beneficial. During the hotter parts of summer, and when the
sun is shining brightly, thousands of these Horse flies appear
on our marshes and inland prairies. There are many different
kinds, over one hundred species of the genus Tabanus alone,
living in North America. Our most common species is the
"Green head," or Tabanus lineola. When about to bite, it set-

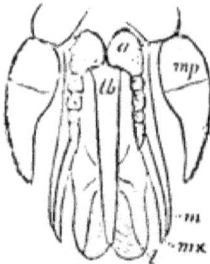

71. Mouth Parts of
Tabanus.

tles quietly down upon the hand, face or
foot, it matters not which, and thrusts its
formidable lancet-like jaws deep into the
flesh. Its bite is very painful, as we can
testify from personal experience. We were
told during the last summer that a horse,
which stood fastened to a tree in a field
near the marshes at Rowley, Mass., was
bitten to death by these Green heads; and
it is known that horses and cattle are occasionally killed by their repeated harassing
bites. In cloudy weather they do not fly, and
they perish on the cool frosty nights of September. The Tumb,
or Tsetze fly, is a species of this group of flies, and while it
does not attack man, plagues to death, and is said to poison by
its bite, the cattle in certain districts of the interior of Africa,
thus almost barring out explorers. On comparing the mouth-
parts of the Horse fly (Fig. 71, mouth of T. lineola), we have
all the parts seen in the mosquito, but greatly modified. Like
the mosquito, the females alone bite, the male Horse fly being
harmless, and frequenting flowers, living upon their sweets.
The labrum (lb), mandibles (m) and maxillæ (mx), are short, stiff
and lancet-like, and the maxillary palpi (mp; a, the five terminal joints of the antennæ) are large, stout, and two-jointed.
While the jaws (both maxillæ and mandibles) are thrust into
the flesh, the tongue (l) spreads around the tube thus formed by
the lancets, and pumps up the blood flowing from the wound, by

aid of the sucking stomach, or crop, being a sac appended to the throat. Other Gad flies, but much smaller, though as annoying to us in woods and fields, are the species of Golden eyed flies, Chrysops, which fly and buzz interminably about our ears, often taking a sudden nip. They plague cattle, settling upon them and drawing their blood at their leisure.

We turn to a comparatively unknown insect, which has occasionally excited some distrust in the minds of housekeepers. It is the carpet fly, Scenopinus pallipes (Fig. 72), which, in the larva state, is found under carpets, on which it is said to feed. The worm (Fig. 73) has a long, white, cylindrical body, divided into twelve segments, exclusive of the head, while the first eight abdominal segments are

72. Carpet Fly.

divided by a transverse suture, so that there appear to be seventeen abdominal segments, the sutures appearing too distinct in the cut. Mr. F. G. Sanborn has reared the fly, here figured, from the worm. The larva also lives in rotten wood; it is too scarce ever to prove very destructive in houses. Either this or a similar fly was once found, we are told by a scientific friend, in great numbers in a "rat" used in dressing a young lady's hair; the worms were living upon the hair stuffing.

73. Carpet Worm.

One of the most puzzling objects to the collector of shells or insects, is the almost spherical larva of Microdon globosus (Fig. 74). It is flattened and smooth beneath and seems to adhere to the under side of stones, where it might be mistaken for a snail.

The Syrphus fly, or Aphis eater, deserves more than the passing notice which we bestow upon it. The maggot (Fig. 75, in the act of devouring an Aphis) is to be sought for established in a group of plant lice (Aphis), which it seizes by means of the long extensible front part of the body. The adult fly (Fig. 76) is gayly spotted and banded with yellow, resembling closely a wasp. It frequents flowers.

The singular rat-tailed pupa-case of Eristalis (Fig. 77) lives in water, and when in want of air, protrudes its long respiratory tube out into the air. We present the figure of an allied fly, Merodon Bardus (Fig. 78; *a*, puparium, natural size). We will

74. Microdon.

75. Syrphus Larva.

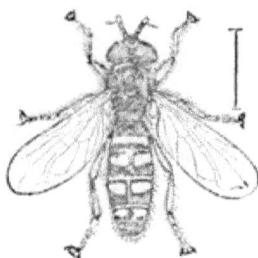

76. Syrphus Fly.

not describe at length the fly, as the admirable drawings of Mr. Emerton cannot fail to render it easily recognizable. The larva is much like the puparium or pupa case, here figured, which closely resembles that of Eristalis, in possessing a long respiratory filament, showing that the maggot undoubtedly lives in the

77. Larva of Rat-tailed Fly.

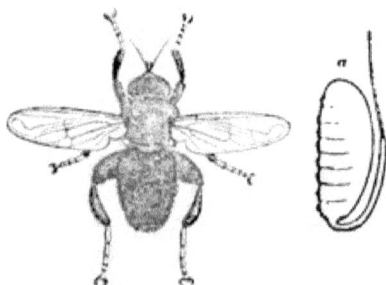

78. Rat-tailed Fly and its Pupa.

water, and when desirous of breathing, protrudes the tube out of the water, thus drawing in air enough to fill its internal respiratory tubes (tracheæ). The Merodon Narcissa probably lives in the soil, or in rotten wood, as the pupa-case has no respiratory tube, having instead a very short, sessile, truncated tube, scarcely as long as it is thick. The case itself is cylindrical, and rounded alike at each end.

We now come to the Bot flies, which are among the most extraordinary, in their habits, of all insects. The history of the Bot flies is in brief thus. The adult two-winged fly lays its eggs on the exterior of the animal to be infested. They are conveyed into the interior of the host, where they hatch, and the worm or maggot lives by sucking in the purulent matter, caused by the irritation set up by its presence in its host; or else the worm itself, after hatching, bores under the skin. When fully grown, it quits the body and finishes its transformations to the fly-state under ground. Many quadrupeds, from mice, squirrels, and rabbits, up to the ox, horse, and even the rhinoceros, suffer from their attacks, while man himself is not exempt. The body of the adult fly is stout and hairy, and it is easily recognized by having the opening of the mouth very small, the mouth-parts being very rudimentary. The larvæ are, in general, thick, fleshy, footless grubs, consisting of eleven seg-

79. Human Bot Worm.

ments, exclusive of the head, which are covered with rows of spines and tubercles, by which they move about within the body,

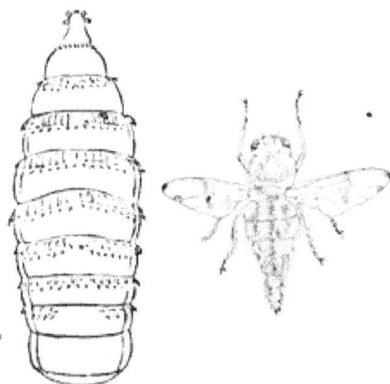

80. Horse Bot Fly.

thus irritating the animals in which they take up their abode. The breathing pores (stigmata) open in a scaly plate at the posterior end of the body. The mouth-parts (mandibles, etc.) of the subcutaneous larvæ consist of fleshy tubercles, while in those species which live in the stomachs and frontal sinuses of their host, they are armed with horny hooks. The larvæ attain their full size after moulting twice.

Just before assuming the pupa state, the maggot leaves its peculiar dwelling place, descends into the ground and there becomes a pupa, though retaining its larval skin, which serves as a protection to it, whence it is called a "puparium."

Several well-authenticated instances are on record of a species of bot fly inhabiting the body of man, in Central and South America, producing painful tumors under the skin of the arm, legs and abdomen. It is still under dispute whether this human bot fly is a true or accidental parasite, the more probable opin-

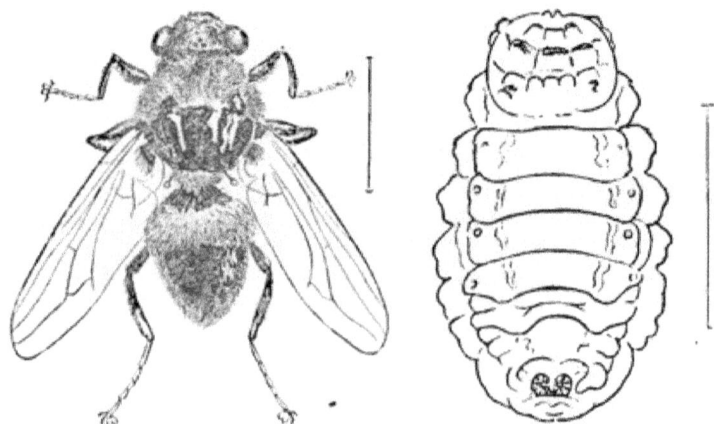

81. Bot Fly of Ox, and Larva.

ion being that its proper host is the monkey or dog. In Cayenne, this revolting grub is called the Ver macaque (Fig. 79); in Para, Ura; in Costa Rica, Torcel; and in New Granada, Gusano peludo, or Nuche. The Dermatobia noxialis, supposed to be the Ver moyocuil of the inhabit-ants of Mexico and New Gra-nada, lives beneath the skin of the dog.

The Bot fly of the horse, (Gastrophilus equi, Fig. 80 and lar-va), is pale yellowish, spotted with red, with short, grayish, yellow

82. Sheep Bot. hairs, and the wings are banded with red-dish. She lays her eggs upon

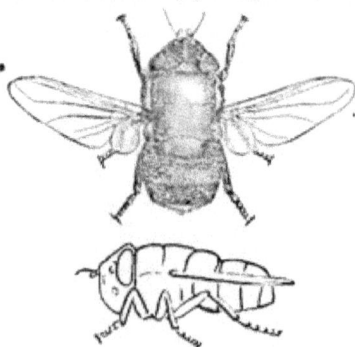

83. Skin Bot Fly.

the knees of the horse. They are conveyed into the stomach, where the larva lives from May until October, and when full grown are found hanging by their mouth hooks on the edge of the rectum of the horse, whence they are carried out in the excrement. The pupa state lasts for thirty or forty days, and the perfect fly appears the next season, from June until October.

The Bot fly of the ox (Hypoderma bovis, Fig. 81, and larva), is black and densely hairy, and the thorax is banded with yellow and white. The larva is found during the month of May, and also in summer, living in tumors on the backs of cattle. When fully grown, which is generally in July, they make their way out and fall to the ground, and live in the pupa-case from twenty-six to thirty days, the fly appearing from May until September. It is found all over the world. The Œstrus ovis, or sheep Bot fly (Fig. 82, larva), is of a dirty ash color. The abdomen is marbled with yellowish and white flecks, and is hairy at the end. This species of Bot fly is larviparous, i. e., the eggs are hatched within the body of the mother, the larvæ being produced alive. M. F. Brauer, of Vienna, the author of the most thorough work we have on these flies, tells me that he knows of but one other Bot fly (a species of Cephanomyia) which produces living larvæ instead of eggs. The eggs of certain other species of Bot flies do not hatch until three or four days after they are laid. The larvæ of the sheep Bot fly live, during April, May and June, in the frontal sinus of the sheep, and also in the nasal cavity, whence they fall to the ground when fully grown. In twenty-four hours they change to pupæ, and the flies appear during the summer.

We also figure the Cuterebra buccata (Fig. 83; a, side view.) which resembles in the larval state the ox Bot fly. Its habits are not known, though the young of other species infest the opossum, squirrel, hare, etc., living in subcutaneous tumors.

The banded Lithacodes.

# CHAPTER VII.

## THE HOUSE FLY AND ITS ALLIES.

THE common House fly, Musca domestica, scarcely needs an introduction to any one of our readers, and its countenance is so well known that we need not present a portrait here. But a study of the proboscis of the fly reveals a wonderful adaptability of the mouth-parts of this insect to their uses. We have already noticed the most perfect condition of these parts as seen in the horse fly. In the proboscis of the house fly the hard parts are obsolete, and instead we have a fleshy tongue-like organ (Fig. 84), bent up beneath the head when at rest. The maxillæ are minute, their palpi (*mp*) being single-jointed, and the mandibles (*m*) are comparatively useless, being very short and small, compared with the lancet-like jaws of the mosquito or horse fly. But the structure of the tongue itself (labium, *l*) is most curious. When the fly settles upon a lump of sugar or other sweet object, it unbends its tongue, extends it, and the broad knob-like end divides into two broad, flat, muscular leaves (*l*), which thus present a sucker-like surface, with which the fly laps up liquid sweets. These two leaves are supported upon a framework of tracheal tubes. In the cut given above, Mr. Emerton has faithfully represented these modified tracheæ, which end in hairs projecting externally.

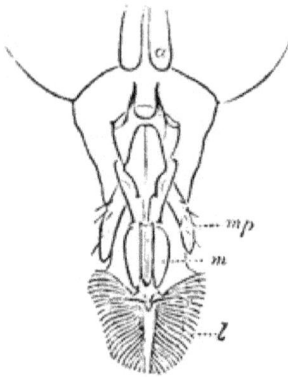

84. Mouth-parts of the House fly.

(80)

Thus the inside of this broad fleshy expansion is rough like a rasp, and as Newport states, " is easily employed by the insect in scraping or tearing delicate surfaces. It is by means of this curious structure that the busy house fly occasions much mischief to the covers of our books, by scraping off the albuminous polish, and leaving tracings of its depredations in the soiled and spotted appearance which it occasions on them. It is by means of these also that it teases us in the heat of summer, when it alights on the hand or face to sip the perspiration as it exudes from, and is condensed upon, the skin."

Every one notices that house flies are most abundant around barns in August and September, and it is in the ordure of stables that the early stages of this insect are passed. No one has traced the transformations of this fly in our country, but we copy from Bouché's work on the transformations of insects, the rather rude figures of the larva (Fig. 85), and pupa-case (*a*) of the Musca domestica of Europe, which is supposed to be our species. Bouché states that the larva is cylindrical, rounded posteriorly, smooth and shining, fleshy, and yellowish white, and four lines long. The pupa-case, or puparium, is dark reddish-brown, and three lines in length. It remains in the pupa state from eight to fourteen days. In Europe it is preyed upon by minute ichneumon flies (Chalcids). The flesh fly, Musca Cæsar, or the Blue-bottle fly, feeds upon decaying animal matter. Its larva (Fig. 86) is long, cylindrical, the head being pointed, and the body conical, the posterior end being squarely docked. The larva of a Sargus-like form which feeds on offal, transforms into a flattened pupa-case (Fig. 87), provided with long, scattered hairs. The House fly disappears in autumn, at the approach of cold weather, though a few individuals pass through the winter, hibernating in houses, and when the rooms are heated may often be seen flying on the windows. Other species fly early in March, on warm days, having hibernated under leaves, and the bark of trees, moss, etc. An allied spe-

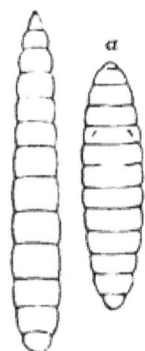

85. Larva; *a*, Pupa-case of House fly.

86. Larva of Flesh fly.

cies, the M. vomitoria, is the Meat fly. Closely allied are the parasitic species of Tachina, which live within the bodies of caterpillars and other insects, and are among the most beneficial of insects, as they prey on thousands of injurious caterpillars.

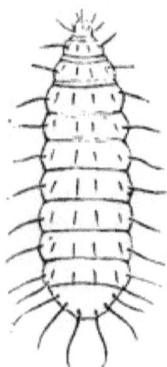

Another fly of this Muscid group, the Idia Bigoti, according to Coquerel and Mondiere, produces in the natives of Senegal, hard, red, fluctuating tumors, in which the larva resides.

Many of the smaller Muscids mine leaves, running galleries within the leaf, or burrowing in seeds or under the bark of plants. We have often noticed blister-like swellings on the bark of the willow, which are occasioned by a cylindrical, short, fleshy larva (Fig. 88 *a*, much enlarged), about a line in length, which changes to a pupa within the old larval skin, assuming the form here represented (Fig. 88*b*), and about the last of June changes to a small black fly

87. Larva of a Sargus-like fly.

(Fig. 88), which Baron Osten Sacken refers doubtfully to the genus Lonchæa.

The Apple midge frequently does great mischief to apples after they are gathered. Mr. F. G. Sanborn states that nine-tenths of the apple crop in Wrentham, Mass., were destroyed by a fly supposed to be the Molobrus mali, or Apple midge, described by Dr. Fitch. "The eggs were supposed to have been laid in fresh apples, in the holes made by the Coddling-moth (Carpocapsa pomo-

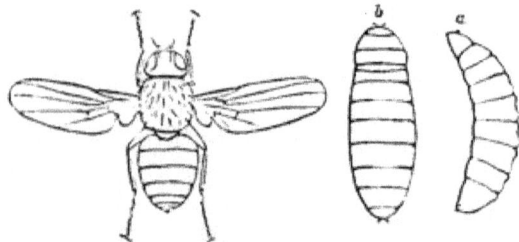

88. Willow Blister fly.

nella), whence the larvæ penetrated into all parts of the apple, working small cylindrical burrows about one-sixteenth of an inch in diameter." Mr. W. C. Fish has also sent me, from Sandwich, Mass., specimens of another kind of apple worm, which he writes has been very common in Barnstable county. "It attacks mostly the earlier varieties, seeming to have a particular fondness for the old fashioned Summer, or High-top

Sweet. The larvæ (Fig. 89 a) enter the fruit usually where it
has been bored by the Apple worm (Carpocapsa), not uncom-
monly through the crescent-like puncture of the curculio, and
sometimes through the calyx, when it has not been troubled by
other insects. Many
of them arrive at ma-
turity in August, and
the fly soon appears,
successive genera-
tions of the maggots
following until cold
weather. I have fre-
quently found the
pupæ in the bottom

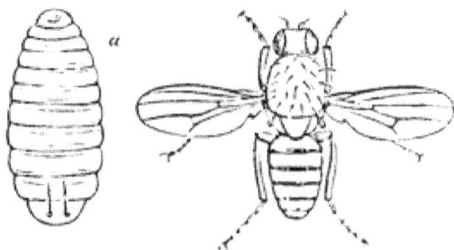

89. Apple Worm and its Larva.

of barrels in a cellar in the winter, and the flies appear in the
spring. In the early apples, the larvæ work about in every
direction. If there be several in an apple, they make it unfit
for use. Apples that appear perfectly sound when taken from
the tree,
will some-
times, if
kept, be
all alive
with them
in a few
weeks."
Baron Os-
ten Sacken
informs
me that it
is a Droso-

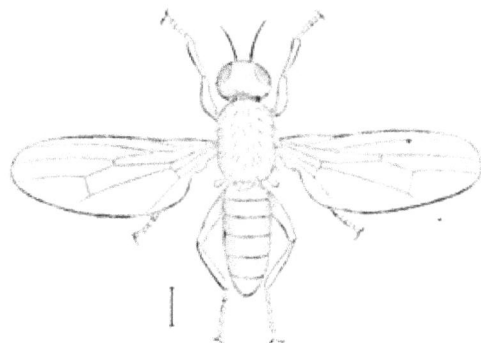

90. Parent of the Cheese Maggot.

91. Pupa-
ca-e of
Wine-fly.

phila, "the species of which live in putrescent vegetable matter,
especially fruits."

An allied fly is the parent of the cheese maggot. The fly
itself (Piophila casei, Fig. 90) is black, with metallic green
reflections, and the legs are dark and paler at the knee-joints,
the middle and hind pair of tarsi being dark honey yellow.
The Wine fly is also a Piophila, and lives the life of a perpetual
toper in old wine casks, and partially emptied beer, cider and
wine bottles, where, with its pupa-case (Fig. 91), it may be
found floating dead in its favorite beverage.

We now come to the more degraded forms of flies which live parasitically on various animals. We figure, from a specimen in the Museum of the Peabody Academy of Science, the Bird tick (Ornithomyia, Fig. 92), which lives upon the Great Horned Owl. Its body is much flattened, adapted for its life under the feathers, where it gorges itself with the blood of its host.

Here belongs also the Horse tick (Hippobosca equina, Fig. 93). It is about the size of the house fly, being black, with yellow spots

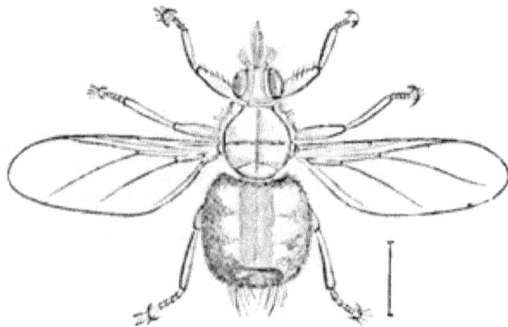

92. Bird Tick.

on the thorax. Verrill* says that "it attacks by preference those parts where the hair is thinnest and the skin softest, especially under the belly and between the hind legs. Its bite causes severe pain, and will irritate the gentlest horses, often rendering them almost unmanageable, and causing them to kick dangerously. When found, they cling so firmly as to be removed with some difficulty, and they are so tough as not to be readily crushed. If one escapes when captured, it will instantly return to the horse, or, perchance, to the

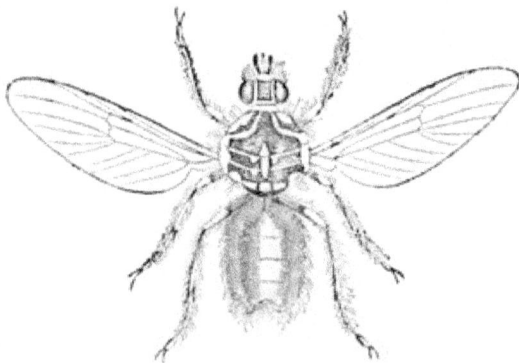

93. The Horse Tick.

* The External and Internal Parasites of Man and Domestic Animals. By Prof. A. E. Verrill, 1870. We are indebted to the author for the use of this and the figures of the Bot fly of the horse, the turkey, duck and hog louse, the cattle tick, the itch insect and mange insect of the horse.

head of its captor, where it is an undesirable guest. Another species sometimes infests the ox."

In the wingless Sheep tick (Melophagus ovinus, Fig. 94, with the pupa-case on the left), the body is wingless and very hairy, and the proboscis is very long. The young are developed within the body of the parent, until they attain the pupa state, when she deposits the pupa-case, which is nearly half as large as her abdomen. Other genera are parasitic on bats; among them are the singular spider-like Bat ticks (Nycteribia, Fig. 95), which have small bodies and enormous legs, and are either blind, or provided with four simple eyes. They are of small size, being only a line or two in length. Such d e g r a d e d

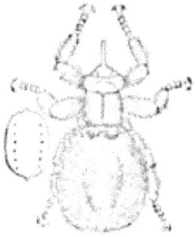

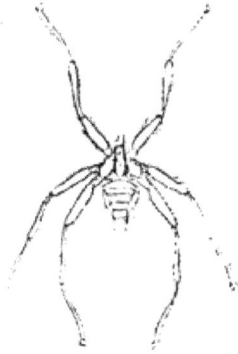

94. Sheep Tick.

95. Bat Tick.

forms of Diptera have a remarkable resemblance to the spiders, mites, ticks, etc. The reader should compare the Nycteribia with the young six-footed moose tick figured farther on. Another spider-like fly is the Chionea valga (Fig. 96; and 97, larva of the European species), which is a degraded Tipula, the latter genus standing near the head of the Diptera. The Chionea, according to Harris. lives in its early stages in the ground like many other gnats, and is found early in the spring, sometimes crawling over the snow. We have also figured and mentioned previously (page 41) the Bee louse, Braula, another wingless spider-like fly.

96. Spider fly.

The Flea is also a wingless fly, and is probably. as 97. Larva of has been suggested by an eminent entomologist, Spider fly. as Baron Osten Sacken informs us, a degraded genus of the family to which Mycetobia belongs. Its transformations are very unlike those of the fly ticks, and agree closely with the

8

early stages of Mycetophila, one of the Tipulid family. In its adult condition the flea combines the characters of the Diptera, with certain features of the grasshoppers and cockroaches, and the bugs. The body of the flea (Fig. 98, greatly magnified; *a*, antennæ; *b*, maxillæ, and their palpi, *c*; *d*, mandibles; the latter, with the labium, which is not shown in the figure, forming the acute beak) is much compressed, and there are minute wing-pads, instead of wings, present in some species.

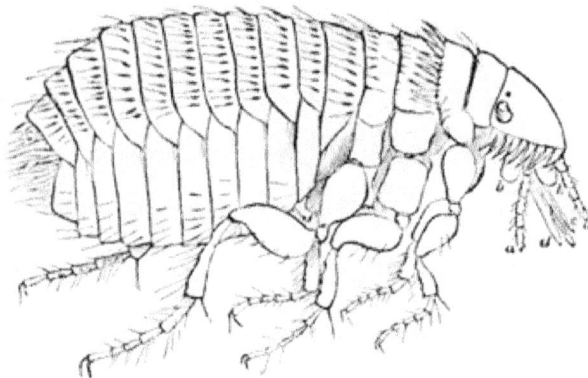

98. Flea, magnified.

Dr. G. A. Perkins, of Salem, has succeeded in rearing in considerable numbers from the eggs, the larvæ of this flea. The larvæ (Fig. 99, much enlarged; *a*, antenna; *b*, the terminal segments of the abdomen), when hatched, are half a line in length. The body is long, cylindrical, and pure white, with thirteen segments exclusive of the head, and provided with rather long hairs. It is very active in its movements, and lives on blood clots, remaining on unswept floors of out-houses, or in the straw or bed of the animals they infest. In six days after the eggs are laid the larvæ appear, and in a few days after leaving the egg they mature, spin a rude cocoon, and change to pupæ, and the perfect insects appear in about ten days.

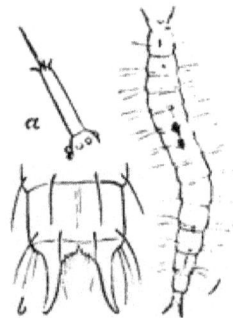

99. Larva of Flea.

A good authority states that the human flea does not exist in America. We never saw a specimen in this country.

A practical point is how to rid dogs of fleas. As a preventive measure, we would suggest the frequent sweeping and cleansing of the floors of their kennels, and renewing the straw or chips composing their beds, — chips being the best material for them to sleep upon. Flea afflicted dogs should be washed every few days in strong soapsuds, or weak tobacco or petroleum water.

A writer in "Science-Gossip" recommends the "use of the Persian Insect Destroyer, one package of which suffices for a good sized dog. The powder should be well rubbed in all over the skin, or the dog, if small, can be put into a bag previously dusted with the powder; in either case the dog should be washed soon after."

One of the most serious insect torments of the tropics of America is the Sarcopsylla penetrans, called by the natives the Jigger, Chigoe, Bicho, Chique, or Pique (Fig. 100, enlarged; *a*, gravid female, natural s i z e ). The female, during the dry season, bores into the feet of the

100. Chigoe.

natives, the operation requiring but a quarter of an hour, usually penetrating under the nails, and lives there until her body becomes distended with eggs, the hind-body swelling out to the size of a pea; her presence often causes distressing sores. The Chigoe lays about sixty eggs, depositing them in a sort of sac on each side of the external opening of the oviduct. The young develop and feed upon the swollen body of the parent flea until they mature, when they leave the body of their host and escape to the ground. The best preventive is cleanliness and the constant wearing of shoes or slippers when in the house, and of boots when out of doors.

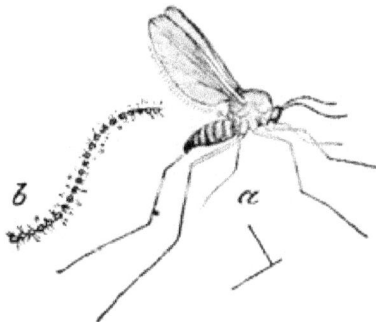

The Willow Gall Fly.

# CHAPTER VIII.

## THE BORERS OF OUR SHADE TREES.

In no way can the good taste and public spirit of our citizens be better shown than in the planting of shade trees. Regarded simply from a commercial point of view one cannot make a more paying investment than setting out an oak, elm, maple or other shade tree about his premises. To a second generation it becomes a precious heirloom, and the planter is duly held in remembrance for those finer qualities of heart and head, and the wise forethought which prompted a deed simple and natural, but a deed too often undone. What an increased value does a fine avenue of shade trees give to real estate in a city? And in the country the single stately elm rising gracefully and benignantly over the wayside cottage. year after year like a guardian angel sending down its blessings of shade, moisture and coolness in times of drought. and shelter from the pitiless storm, recalls the tenderest associations of generation after generation that go from the old homestead.

Occasionally the tree, or a number of them, sicken and die, or linger out a miserable existence. and we naturally after failing to ascribe the cause to bad soil, want of moisture or adverse atmospheric agencies, conclude that the tree is infested with insects, especially if the bark in certain places seems diseased. Often the disease is in streets lighted by gas, attributed to the leakage of the gas. Such a case has come up recently at Morristown, New Jersey. An elm was killed by the Elm borer (Compsidea tridentata), and the owner was on the point of suing the Gas Company for the loss of the tree from the supposed leakage of a gas pipe. While the matter was in dispute, a gentleman of that city took the pains to peel off a piece of the

(88)

bark and found, as he wrote me, "great numbers of the larvæ
of this beetle in the bark and between the bark and the wood,
while the latter is 'tattooed' with sinuous grooves in every
direction and the tree is completely girdled by them in some
places. There are three different sizes of the larvæ, evidently
one, two and three years old, or more properly six, eighteen and
thirty months old." The tree had to be cut down.

Dr. Harris, in his "Treatise on Injurious Insects," gives an
account of the ravages of this insect, which we quote: "On
the 19th of June, 1846, Theophilus Parsons, Esq., sent me some
fragments of bark and insects which were taken by Mr. J. Rich-
ardson from the decaying elms on Boston Common, and among
the insects I recognized a pair of these beetles in a living state.
The trees were found to have suffered terribly from the ravages
of these insects. Several of them had already been cut down,
as past recovery; others were in a dying state, and nearly all
of them were more or less affected with disease or premature
decay. Their bark was perforated, to the height of thirty feet
from the ground, with numerous holes, through which insects
had escaped; and large pieces had become so loose, by the
undermining of the grubs, as to yield to slight efforts, and come
off in flakes. The inner bark was filled with burrows of the
grubs, great numbers of which, in various stages of growth,
together with some in the pupa state, were found therein; and
even the surface of the wood, in many cases, was furrowed with
their irregular tracks. Very rarely did they seem to have penetrated far into the wood itself; but their operations were mostly
confined to the inner layers of the bark, which thereby became
loosened from the wood beneath. The grubs rarely exceed
three-quarters of an inch in length. They have no feet, and
they resemble the larvæ of other species of Saperda, except
in being rather more flattened. They appear to complete their
transformations in the third year of their existence.

"The beetles probably leave their holes in the bark during
the month of June and in the beginning of July, for, in the
course of thirty years, I have repeatedly taken them at various
dates, from the fifth of June to the tenth of July. It is evident,
from the nature and extent of their depredations, that these
insects have alarmingly hastened the decay of the elm trees on
Boston Mall and Common, and that they now threaten their
entire destruction. Other causes, however, have probably con-

tributed to the same end. It will be remembered that these trees have greatly suffered, in past times, from the ravages of canker-worms. Moreover, the impenetrable state of the surface soil, the exhausted condition of the subsoil, and the deprivation of all benefit from the decomposition of accumulated leaves, which, in a state of nature, the trees would have enjoyed, but which a regard for neatness has industriously removed, have doubtless had no small influence in diminishing the vigor of the trees, and thus made them fall unresistingly a prey to insect-devourers. The plan of this work precludes a more full consideration of these and other topics connected with the growth and decay of these trees; and I can only add, that it may be prudent to cut down and burn all that are much infested by the borers."

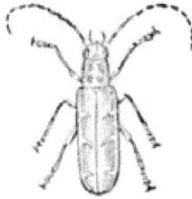

The Three-toothed Compsidea (Fig. 101), is a rather flat-bodied, dark brown beetle,

101. Elm Tree Beetle.

with a rusty red curved line behind the eyes, two stripes on the thorax, and a three-toothed stripe on the outer edge of each wing cover. It is about one-half an inch in length.

The larva (Fig. 102) is white, subcylindrical, a little flattened, with the lateral fold of the body rather prominent; the end of the body is flattened, obtuse, and nearly as wide at the end as at the first abdominal ring. The head is one-half as wide as the prothoracic ring, being rather large. The prothoracic ring, or segment just behind the head, is transversely oblong, being twice as broad as long; there is a pale dorsal corneous transversely oblong shield, being about two-thirds as long as wide, and nearly as long as the four succeeding segments; this plate is smooth, except on the posterior half, which is rough, with the front edge irregular and not extending far down the sides. Fine hairs arise from the front edge and side of

102. Elm Tree Borer.

the plate, and similar hairs are scattered over the body and especially around the end. On the upper side of each segment is a transversely oblong ovate roughened area, with the front edge slightly convex, and the hinder slightly arcuate. On the under side of each segment are similar rough horny plates, but arcuate in front, with the hinder edge straight.

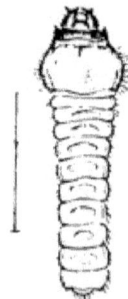

It differs from the larva of the Linden tree borer (Saperda vestita) in the body being shorter, broader, more hairy, with the tip of the abdomen flatter and more hairy. The prothoracic segment is broader and flatter, and the rough portion of the dorsal plates is larger and less tranversely ovate. The structure of the head shows that its generic distinctness from Saperda is well founded, as the head is smaller and flatter, the clypeus being twice as large, and the labrum broad and short, while in S. vestita it is longer than broad. The mandibles are much longer and slenderer, and the antennæ are much smaller than in S. vestita.

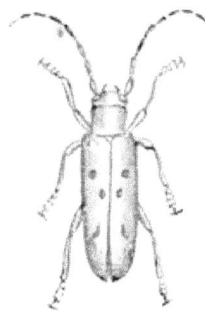

103. Linden Tree Beetle.

The Linden tree borer (Fig. 103) is a greenish snuff-yellow beetle, with six black spots near the middle of the back; and it is about eight-tenths of an inch in length, though often smaller. The beetles, according to Dr. Paul Swift, as quoted by Dr. Harris, were found (in Philadelphia) upon the small branches and leaves on the 25th day of May, and it is said that they come out as early as the first of the month, and continue to make their way through the back of the trunk and large branches during the whole of the warm season. They immediately fly into the top of the tree, and there feed upon the epidermis of the tender twigs, and the petioles of the

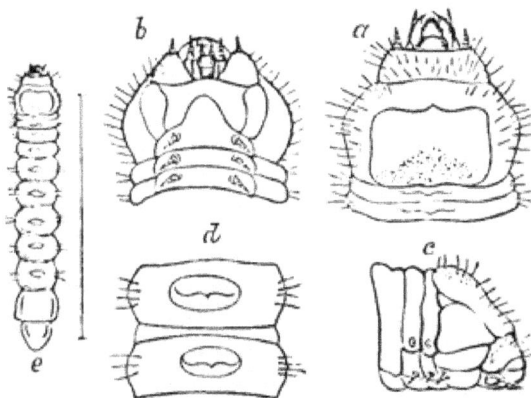

104. Linden Tree Borer.

leaves, often wholly denuding the latter, and causing the leaves to fall. They deposit their eggs, two or three in a place, upon the trunk or branches especially about the forks, making slight incisions or punctures for their reception with their strong

jaws. As many as ninety eggs have been taken from a single
beetle. The grubs (Fig. 104, e; a, enlarged view of the head
seen from above; b, the under view of the same; c, side view,
and d, two rings of the body enlarged), hatched from these
eggs, undermine the bark to the
extent of six or eight inches, in
sinuous channels, or penetrate the
solid wood an equal distance. It
is supposed that three years are
required to mature the insect.
Various expedients have been
tried to arrest their course, but
without effect. A stream, thrown
into the tops of trees from the
hydrant, is often used with good
success to dislodge other insects;
but the borer-beetles, when thus
disturbed, take wing and hover
over the trees till all is quiet, and

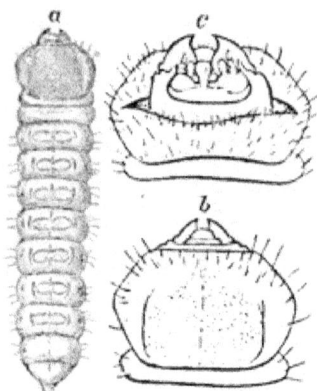

105. Poplar Tree Borer.

then alight and go to work again. The trunks and branches
of some of the trees have been washed over with various prep-
arations without benefit. Boring the trunk near the ground
and putting in
sulphur and
other drugs, and
plugging, have
been tried with
as little effect.

The city of
Philadelphia has
suffered griev-
ously from this
borer.

Dr. Swift re-
marks, in 1844,
that "the trees
in Washington
and Indepen-
dence Squares

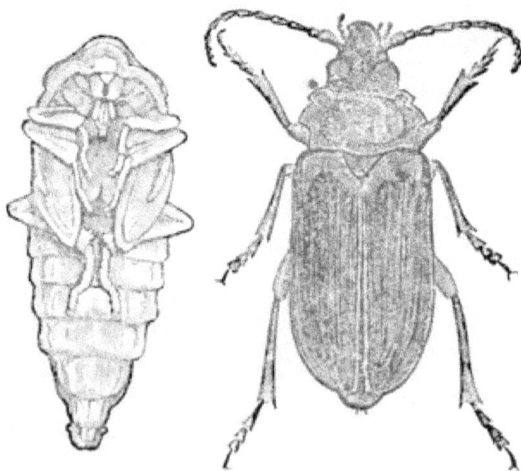

106. Broad-necked Prionus.

were first observed to have been attacked about seven years
ago. Within two years it has been found necessary to cut down

forty-seven European lindens in the former square alone, where
there now remain only a few American lindens, and these
a good deal eaten." In New England this beetle should be
looked for during
the first half of
June.

The Poplar tree
is infested by an-
other species of Sa-
perda (S. calcarata).
This is a much larger
beetle than those
above mentioned,
being an inch or a
little more in length.
It is gray, irregu-
larly striped with

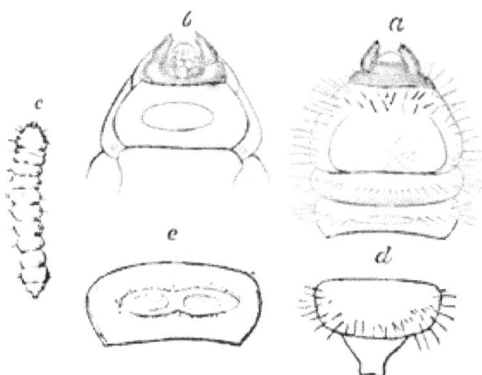

107. Larva of the Plain Saperda.

ochre, and the wing-covers end in a sharp point. The grub
(Fig. 105 a; b, top view of the head; c, under side) is about
two inches long and whitish yellow. It has, with that of the
Broad-necked Prionus (P. laticollis of Drury, Fig. 106, adult
and pupa), as Harris states, "almost entirely destroyed the

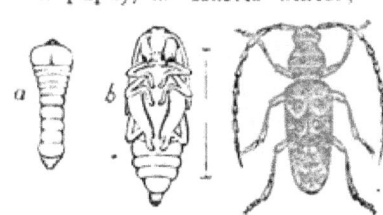

108. Locust Borer.

Lombardy poplar in this vicinity" (Boston). It bores in the
trunks, and the beetle flies by
night in August and September. We also figure the larva
of another borer (Fig. 107 c;
a, top view of the head; b,
under side; c, dorsal view of
an abdominal segment; d, end
of the body, showing its peculiar form), the Saperda inornata of
Say, the beetle of which is black, with ash gray hairs, and without spines on the wing-covers. It is much smaller than any of
the foregoing species, being nine-twentieths of an inch in length.
Its habits are not known. We also figure the Locust and Hick-
ory borer (Fig. 108; a, larva; b, pupa), which has swept off the
locust tree from New England. The beautiful yellow banded
beetles are very abundant on the flowers of the golden rod in
September.

# CHAPTER IX.

## CERTAIN PARASITIC INSECTS.

The subject of our discourse is not only a disagreeable but too often a painful one. Not only is the mere mention of the creature's name of which we are to speak tabooed and avoided by the refined and polite, but the creature itself has become extinct and banished from the society of the good and respectable. Indeed under such happy auspices do a large proportion of the civilized world now live that their knowledge of the habits and form of a louse may be represented by a blank. Not so with some of their great-great-grandfathers and grandmothers, if history, sacred and profane, poetry,* and the annals of literature testify aright; for it is comparatively a recent fact in history that the louse has awakened to find himself an outcast and an alien. Among savage nations of all climes, some of which have been dignified with the apt, though high sounding name of Phthiriophagi, and among the Chinese and other semi-civilized peoples, these lords of the soil still flourish with a luxuriance and rankness of growth that never diminishes, so that we may say without exaggeration that certain mental traits and fleshly appetites

---

* Ha! whare ye gaun, ye crowlin ferlie!
Your impudence protects you sairly:
I canna say but ye strunt rarely,
    Owre gauze and lace;
Tho' faith, I fear ye dine but sparely
    On sic a place.

Ye ugly, creepin, blastie wonner,
Detested, shunn'd by saunt and sinner,
How dare ye set your fit upon her
    Sae fine a lady!
Gae somewhere else and seek your dinner
    On some poor body.

             (To a Louse. — Burns.)

(94)

induced by their consumption as an article of food may have
been created, while a separate niche in our anthropological mu-
seums is reserved for the instruments of warfare, both offensive
and defensive, used by their phthiriophagous hunters. Then have
we not in the very centres of civilization the poor and degraded,
which are most faithfully attended by these revolting satellites!

But bantering aside, there is no more engaging subject to the
naturalist than that of animal parasites. Consider the great
proportion of animals that gain their livelihood by stealing that
of others. While a large proportion of plants are more or less
parasitic, they gain thereby in interest to the botanist, and
many of them are eagerly sought as the choicest ornaments of
our conservatories. Not so with their zoological confrères.
All that is repulsive and uncanny is associated with them, and
those who study them, though perhaps among the keenest intel-
lects and most industrious observers, speak of them without
the limits of their own circle in subdued whispers or under a
protest, and their works fall under the eyes of the scantiest
few. But the study of animal parasites has opened up new
fields of research, all bearing most intimately on those two
questions that ever incite the naturalist to the most laborious
and untiring diligence — what is life and its origin? The subjects of the alternation of generations, or parthenogenesis, of
embryology and biology, owe their great advance, in large
degree, to the study of such animals as are parasitic, and the
question whether the origin of species be due to creation by
the action of secondary laws or not, will be largely met and
answered by the study of the varied metamorphoses and modes
of growth, the peculiar modification of organs that adapt them
to their strange modes of life, and the consequent variation in
specific characters so remarkably characteristic of those animals living parasitically upon others. *

With these considerations in view surely a serious, thoughtful, and thorough study of the louse, in all its varieties and
species, is neither belittling nor degrading, nor a waste of time.
We venture to say, moreover, that more light will be thrown
on the classification and morphology of insects by the study of

---

* We notice while preparing this article that a journal of Parasitology has for
some time been issued in Germany — that favored land of specialists. It is the
"Zeitschrift für Parasitenkunde," edited by Dr. E. Hallier and F. A. Zurn. 8vo,
Jena.

the parasitic species, and other degraded, wingless forms that do not always live parasitically, especially of their embryology and changes after leaving the egg, than by years of study of the more highly developed insects alone. Among Hymenoptera the study of the minute Ichneumons, such as the Proctotrupids and Chalcids, especially the egg-parasites; among moths the study of the wingless canker-worm moth and Orgyia; among Diptera the flea, bee louse, sheep tick, bat tick, and other wingless flies; among Coleoptera, the Meloë, and singular Stylops and Xenos; among Neuroptera, the snow insect, Boreus, the Podura (Fig. 109) and Lepisma, and especially the hemipterous lice, will throw a flood of light on these prime subjects in philosophical entomology.

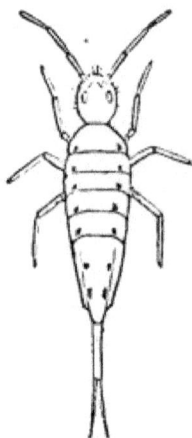

Without farther apology, then, and very dependent on the labors of others for our information, we will say a few words on some interesting points in the natural history of lice. In the first place, how does the louse bite? It is the general opinion among physicians, supported by able entomologists, that the louse has jaws, and bites. But while the bird lice (Mallophaga) do have biting jaws, whence the Germans call them skin-eaters (*pelzfresser*), the mouth parts of the genus Pediculus, or true louse, resemble in their structure those of the bed-bug (Fig. 110), and other Hemiptera. In its form the louse closely resembles the bed-bug, and the two groups of lice, the Pediculi and Mallophaga, should be considered as families of Hemiptera, though degraded and at the base of the hemipterous series. The resemblance is carried out in the form of the egg, the mode of growth of the embryo, and the metamorphosis of the insect after leaving its egg.

109. Podura.

110. Bed-bug.

Schiödte, a Danish entomologist, has, it seems to us, forever settled the question as to whether the louse bites the flesh or sucks blood, and decides a point interesting to physicians, i. e., that the loathsome disease called phthiriasis is a nonentity. From this source not only many living in poverty and squalor are said to have died, but also men of renown, among whom Denny in his work on the Anoplura, or lice, of Great Britain, mentions the name of "Pheretima, as recorded by Herodotus, Antiochus Epiphanes, the Dictator Sylla, the two Herods, the Emperor Maximian, and Phillip the Second." Schiödte, in his essay "On Phthirius, and on the Structure of the Mouth in Pediculus" (Annals and Magazine of Natural History, 1866, page 213), says that these statements will not bear examination, and that this disease should be placed on the "retired list," for such a malady is impossible to be produced by simply blood-sucking animals, and that they are only the disgusting attendants on other diseases. Our author thus describes the mouth parts of the louse.

"Lice are no doubt to be regarded as bugs, simplified in structure and lowered in animal life in accordance with their mode of living as parasites, being small, flattened, apterous, myopic, crawling and climbing, with a conical head, moulded as it were to suit the rugosities of the surface they inhabit, provided with a soft, transversely furrowed skin, probably endowed with an acute sense of feeling, which can guide them in that twilight in which their mode of life places them. The peculiar attenuation of the head in front of the antennæ at once suggests to the practised eye the existence of a mouth adapted for suction. This mouth differs from that of the Hemiptera (bed-bug, etc.) generally, in the circumstance that the labium is capable of being retracted into the upper part of the head, which therefore presents a little fold, which is extended when the labium is protruded. In order to strengthen this part, a flat band of chitine is placed on the under surface, just as the shoemaker puts a small piece of gutta-percha into the back of an India-rubber shoe; as, however, the chitine is not very elastic, this band is rather thinner in the middle, in order that it may bend and fold a little when the skin is not extended by the lower lip. The latter consists, as usual, of two hard lateral pieces, of which the fore ends are united by a membrane so that they form a tube, of which the interior covering is a continuation of the elastic membrane in the top of the head; inside its orifice there are a number of small hooks, which assume different positions according to the degree of protrusion; if this is at its highest point the orifice is turned inside out, like a collar, whereby the small hooks are directed backwards, so that they can serve as barbs. These are the movements which the animal executes after having first inserted the labium through a sweat-pore. When the hooks have got a firm hold, the first pair of setæ (the real mandibles transformed) are protruded; these are, towards their points, united by a membrane so as to form a closed tube, from which, again, is inserted the second pair of setæ, or maxillæ, which in the same manner are transformed into a tube ending in four small lobes placed crosswise. It follows that when the whole instrument is exserted, we perceive a long membranous flexible tube hanging down from the labium, and along the walls of this tube the setiform mandibles and maxillæ in the shape of long narrow bands of chitine. In this way the tube of suction can be made longer or shorter as required, and easily adjusted to the thickness of the skin in the particular place where the animal is sucking, whereby access to the capillary system is secured at any part of the body. It is apparent, from the whole structure of the instrument, that it is by no means calculated on being used as a sting, but is rather to be compared to a delicate elastic probe, in the use of which the terminal lobes probably serve as feelers. As soon as the capillary system is reached, the blood will at once ascend into the narrow tube, after which the current is continued with increasing rapidity by means of the pulsation of the pumping ventricle and the powerful peristaltic movement of the digestive tube."

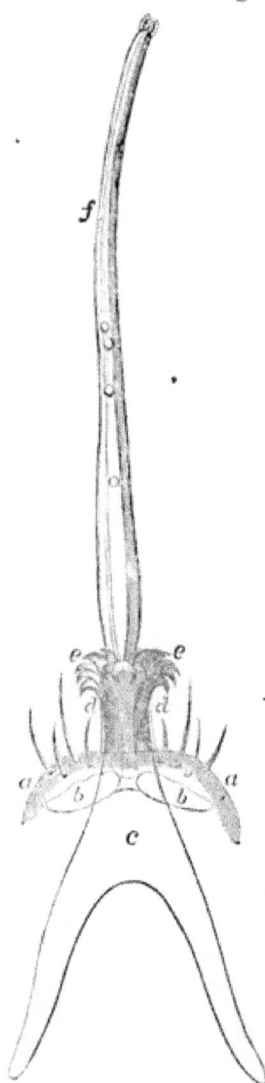

111. Mouth of the Louse.

If we compare the form of the louse (Fig. 112, Pediculus

* Figure 111 represents the parts of the mouth in a large specimen of *Pediculus vestimenti*, entirely protruding, and seen from above, magnified one hundred and sixty times; *aa*, the summit of the head with four bristles on each side; *bb*, the

capitis, the head louse; Fig. 113, P. vestimenti, the body louse)
with the young bed-bug as figured by Westwood (Modern
Classification of Insects, ii, p. 475) we shall see a very close
resemblance, the head of the young Cimex being proportionally
larger than in the adult, while the thorax is smaller, and the
abdomen is more ovate, less rounded; moreover the body is
white and partially transparent.

Under a high power of the microscope specimens treated
with diluted potash show that the mandibles and
maxillæ arise near each other in the middle of the
head opposite the eyes, their bases slightly diverging. Thence they converge to the mouth, over
which they meet, and beyond are free, being hollow, thin bands of chitine, meeting like the maxillæ,
or tongue, of butterflies to form a hollow tube for
suction. The mandibles each suddenly end in a
curved, slender filament, which is probably used
as a tactile organ to explore the best sites in the
flesh of their victim for drawing blood. On the
other hand the maxillæ, which are much narrower

112. Head
Louse.

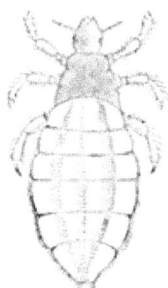

than the mandibles, become rounded towards the end, bristle-
like, and tipped with numerous exceedingly fine barbs, by which
the bug anchors itself in the flesh, while the
blood is pumped through the mandibles. The
base of the large, tubular labium, or beak, which
ensheathes the mandibles and maxillæ, is opposite the end of the clypeus or front edge of the
upper side of the head, and at a distance beyond
the mouth equal to the breadth of the labium
itself. The labium, which is divided into three
joints, becomes flattened towards the tip, which
is square, and ends in two thin membranous

113. Body Louse. lobes, probably endowed with a slight sense of
touch. On comparing these parts with those of the louse, it
will be seen how much alike they are with the exception of the
labium, a very variable organ in the Hemiptera. From the long

chitinous band, and c, the hind part of the lower lip, such as they appear through
the skin by strong transmitted light; dd, the foremost protruding part of the lower
lip (the haustellum); ee, the hooks turned outwards; f, the inner tube of suction,
slightly bent and twisted; the two pairs of jaws are perceived on the outside as
thin lines; a few blood globules are seen in the interior of the tube.

sucker of the Pediculus, to the stout chitinous jaws of the Mallophaga, or bird lice, is a sudden transition, but on comparing the rest of the head and body it will be seen that the distinction only amounts to a family one, though Burmeister placed the Mallophaga among the Orthoptera (grasshoppers and crickets) on account of the mandibles being adapted for biting. It has been a common source of error to depend too much upon one or a single set of organs. Insects have been classified on characters drawn from the wings, or the number of the joints of the tarsi, or the form of the mouth parts. We must take into account in endeavoring to ascertain the limits of natural groups, all the organs collectively, as well as the internal anatomy and the embryology and metamorphosis of insects, before we can hope to obtain a natural classification.

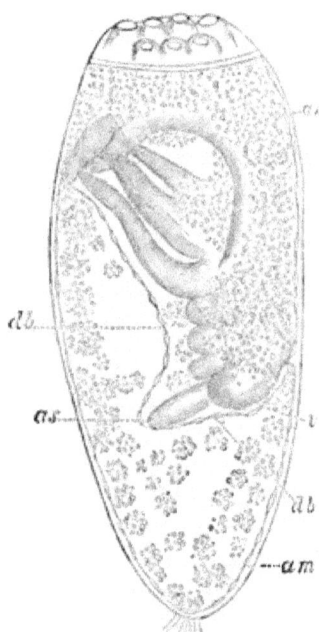

114. Embryo of the Louse.

The family of bird lice is a very extensive one, embracing many genera, and several hundred species. One or more species infest the skin of all our domestic and wild mammals and birds, some birds sheltering beneath their feathers four or five species of lice. Before giving a hasty account of some of our more common species, we will give a sketch of the embryological history of the lice, with special reference to the structure of the mouth parts.

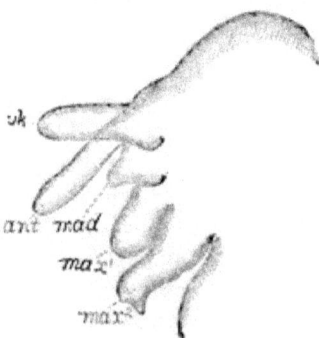

115. Mouth Parts of the Louse.

The eggs (Fig. 114, egg of the head louse) are long, oval, somewhat pear-shaped, with the hinder end somewhat pointed, while the anterior end is flattened, and bears little conical micropyles (*m*, minute orifices for the passage of the spermatozoa into the egg), which vary in form in the different species and genera; the opposite end of the egg is provided with a few bristles. The female attaches her eggs to the hairs or feathers of her host.

After the egg has been fertilized by the male, the blastoderm, or primitive skin, forms, and subsequently two layers, or em-bryonal membranes, ap-

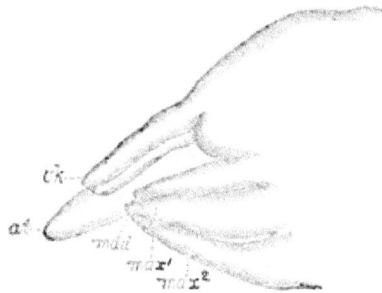

116. Mouth Parts of the Louse.

pear; the outer is called the amnion (Fig. 114, *am*), while the inner visceral membrane (*db*) partially wraps the rude form of the embryo in its folds. The head (*vk*) of the embryo is now directed towards the end of the egg on which the hairs are situated; afterwards the embryo revolves on its axis and the head lies next to the opposite end of the egg. Eight tubercles bud out from the under side of the head, of which the foremost and longest are the antennæ (*as*), those succeeding are the mandi-

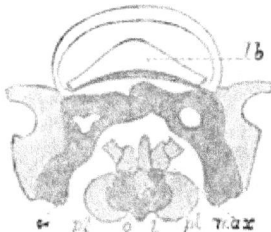

118. Mouth Parts of Louse.

bles, maxillæ, and second maxillæ, or labium. Behind them arise six long, slender tubercles forming the legs, and the primitive streak rudely marks the lower wall of the thorax and abdomen not yet formed. Figure 115 represents the head and mouth parts of the embryo of the same louse; *rk* is the forehead, or clypeus; *ant*, the anten-næ; *mad*, the mandibles;

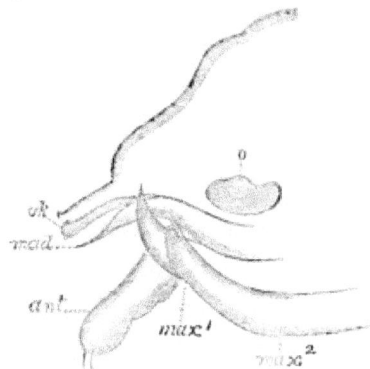

117. Mouth Parts of Louse.

*max*[1], the first pair of maxillæ, and *max*[2], the second pair of maxillæ, or ·labium. Figure 116 represents the mouth parts of the same insect a little farther advanced, with the jaws and labium elongated and closely folded together. Figure 117 represents the same still farther advanced; the mandibles (*mad*) are

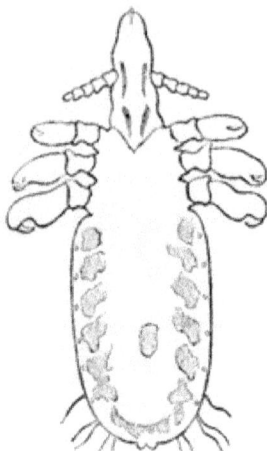

119. Louse of Cow.

sharp, and resemble the jaws of the Mallophaga; and the maxillæ (*max¹*) and labium (*max²*) are still large, while afterwards the labium becomes nearly obsolete. Figure 118 represents a front view of the mouth parts of a bird louse, G o n i o d e s ; *lb*, is the upper lip, or labrum, lying under the clypeus; *mad*, the mandibles; *max*, the maxillæ; *l*, the l y r e - f o r m e d piece; and *pl*, the "plate."

We will now describe some of the common species of lice found on a few of our domestic animals, and the mallophagous parasites occurring on certain mammals and birds. The family Pediculina, or true lice, is higher than the bird lice, their mouth parts, as well as the structure of the head, resembling the true Hemiptera, especially the bed bug. The clypeus, or front of the head, is much smaller than in the bird lice, the latter retaining the enlarged forehead of the embryo, it being in some species half as large as the rest of the head.

All of our domestic mammals and birds are plagued by one or more species of lice. Figure 119 represents the Hæmatopinus vituli, which is brownish in color. As the specimen figured came from the Burnett collection of the Boston Society of Natural History, together with those of the goat louse, the louse of the common fowl, and of the cat, they are undoubtedly naturalized here.

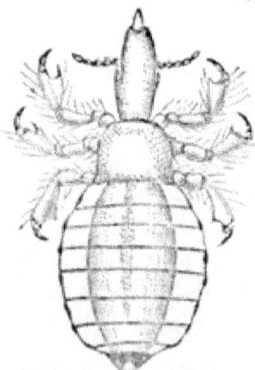

120. Louse of Hog.

Quite a different species is the louse of the hog (H. suis, Fig. 120).

The remaining parasites belong to the skin-biting lice, or Mallophaga, and I will speak of the several genera referred to in their natural order, beginning with the highest form and that which is nearest allied to Pediculus.

The common barn-yard fowl is infested by a louse that we have called Goniocotes Burnettii (Fig. 121), in honor of the late Dr. W. I. Burnett, a young and talented naturalist and physiologist, who paid more attention than any one else in this country to the study of these parasites, and made a large collection of them, now in the museum of the Boston Society of Natural History. It differs from the G. hologaster of Europe, which lives on the same bird, in the short second joint of the antennæ, which are also stouter; and in the long head, the clypeus being much longer and more acutely rounded; while the head is less hollowed out at the insertion of the antennæ.

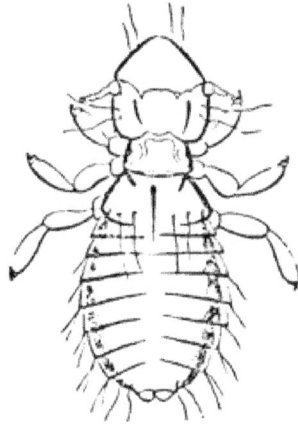

121. Louse of Domestic Fowl.

The abdomen is oval, and one-half as wide as long, with transverse, broad, irregular bands along the edges of the segments. The mandibles are short and straight, two toothed. The body is slightly yellowish, and variously streaked and banded with pitchy black. The duck is infested by a remarkably slender form (Fig. 122, Philopterus squalidus). Figure 123 represents the louse of the cat, and another species (Fig. 124) of the same genus (Trichodes) lives upon the goat.

The most degraded genus is Gyropus. Mr. C. Cook has found Gyropus ovalis of Europe abundant on the Guinea pig. A species is also found on the porpoise; an interesting fact, as this is the only insect we know of that lives parasitically on any marine animal.

122. Duck Louse.

The genus Goniodes (Fig. 125, G. stylifer, the turkey louse) is of great interest from a morphological and developmental point of view, as the antennæ are described and figured by Denny as being " in the males cheliform (Fig. 126, a, male; b, female); the first joint being very large and thick, the third considerably smaller, recurved towards the first, and forming a claw, the fourth and fifth very small, arising from the back of the third." He farther remarks, that "the males of this [which lives on the turkey] and all the other species of Goniodes, use the first and third joints of the antennæ with great facility, acting the part of a finger and thumb."

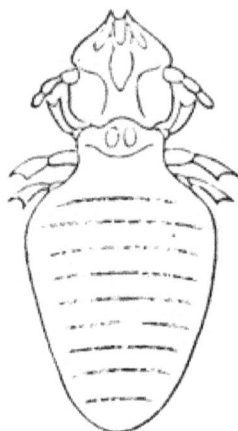

123. Louse of the Cat.

The antennæ of the females are of the ordinary form. This hand-like structure, is, so far as we know, without a parallel among insects, the antennæ of the Hemiptera being almost uniformly filiform, and from two to nine-jointed. The design of this structure is probably to enable the male to grasp its consort and also perhaps to cling to the feathers, and thus give it a superiority over the weaker sex in its advances towards courtship. Why is this advantage possessed by the males of this genus alone? The world of insects, and of animals generally abounds in such instances, though existing in other organs, and the developmentist dimly perceives in such departures from a normal type of structure, the origin of new generic forms, whether due at first to a seemingly accidental variation, or, as in this instance, perhaps, to long use as prehensile organs through successive generations of lice having the antennæ slightly diverging from the typical condition, until the present form has been developed. Another generation of naturalists will perhaps unanimously agree that the Creator has thus worked through secondary laws, which many of the naturalists of the present day are endeavoring, in a truly scientific and honest spirit of inquiry, to discover.

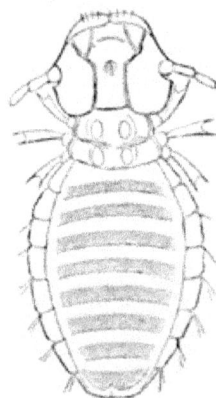

124. Louse of the Goat.

In their claw or leg-like form these male antennæ also repeat in the head, the general form of the legs, whose prehensile and grasping functions they assume. We have seen above that the appendages of the head and thorax are alike in the embryo, and the present case is an interestirg example of the unity of type of the jointed appendages of insects, and articulates generally.

Another point of interest in these degraded insects is, that the process of degradation begins either late in the life of the embryo or during the changes from the larval to the adult, or winged state. An instance of the latter may be observed in the wingless female of the canker worm, so different from the winged male; this difference is created after the larval stage, for the caterpillars of both sexes are the same, so far as we know. So with numerous other examples among the moths. In the louse, the embryo, late in its life, resembles the embryos of other insects, even Corixa, a member of a not

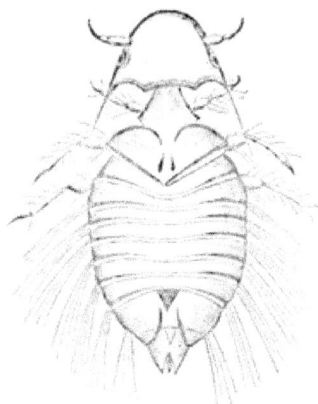

125. The Turkey Louse.

126. Antennæ of Goniodes.

remotely allied family. But just before hatching the insect assumes its degraded louse physiognomy. The developmentist would say that this process of degradation points to causes acting upon the insect just before or immediately after birth, inducing the retrogression and retardation of development, and would consider it as an argument for the evolution of specific forms by causes acting on the animal while battling with its fellows in the struggle for existence, and perhaps consider that the metamorphoses of the animal within the egg are due to a reflex action of the modes of life of the ancestors of the animal on the embryos of its descendants.

# CHAPTER X.

## THE DRAGON FLY.

WERE we to select from among the insects a type of all that is savage, relentless, and bloodthirsty, the Dragon fly would be our choice. From the moment of its birth until its death, usually a twelve-month, it riots in bloodshed and carnage. Living beneath the waters perhaps eleven months of its life, in the larva and pupa states, it is literally a walking pitfall for luckless aquatic insects; but when transformed into a fly, ever on the wing in pursuit of its prey, it throws off all concealment, and reveals the more unblushingly its rapacious character.

Not only do its horrid visage and ferocious bearing frighten children, who call it the "Devil's Darning-needle," but it even distresses older persons, so that its name has become a byword. Could we understand the language of insects, what tales of horror would be revealed! What traditions, sagas, fables, and myths must adorn the annals of animal life regarding this Dragon among insects!

To man, however, aside from its bad name and its repulsive aspect, which its gay trappings do not conceal, its whole life is beneficent. It is a scavenger, being like that class ugly and repulsive, and holding literally, among insects, the lowest rank in society. In the water, it preys upon young mosquitoes and the larvæ of other noxious insects. It thus aids in maintaining the balance of life, and cleanses the swamps of miasmata, thus purifying the air we breathe. During its existence of three or four weeks above the waters, its whole life is a continued good to man. It hawks over pools and fields and through gardens, decimating swarms of mosquitoes, flies, gnats, and other baneful insects. It is a true Malthus' delight, and, following that sanguinary philosopher, we may believe that our Dragon fly is an

(106)

entomological Tamerlane or Napoleon sent into the world by a
kind Providence to prevent too close a jostling among the myriads of insect life.

We will, then, conquer our repugnance to its ugly looks and
savage mien, and contemplate the hideous monstrosity,— as it
is useless to deny that it combines the graces of the Hunchback
of Notre Dame and Dickens' Quilp, with certain features of its
own,— for the good it does in Nature.

Even among insects, a class replete with forms the very incarnation of ugliness and the perfection of all that is hideous in
nature, our Dragon fly is most conspicuous. Look at its enormous head, with its beetling brows, retreating face, and heavy
under jaws,—all eyes and teeth,—and hung so loosely on its
short, weak neck, sunk beneath its enormous hunchback,—for
it is wofully round-shouldered, — while its long, thin legs,
shrunken as if from disease, are drawn up beneath its breast,
and what a hobgoblin it is!

Its gleaming wings are, however, beautiful objects. They
form a broad expanse of delicate parchment-like membrane
drawn over an intricate network of veins. Though the body is
bulky, it is yet light, and easily sustained by the wings. The
long tail undoubtedly acts as a rudder to steady its flight.

These insects are almost universally dressed in the gayest
colors. The body is variously banded with rich shades of blue,
green, and yellow, and the wings give off the most beautiful
iridescent and metallic reflections.

During July and August the various species of Libellula and
its allies most abound. The eggs are attached loosely in bunches
to the stems of rushes and other water-plants. In laying them,
the Dragon fly, according to Mr. P. R. Uhler's observations,
"alights upon water-plants, and, pushing the end of her body
below the surface of the water, glues a bunch of eggs to the
submerged stem or leaf. Libellula auripennis I have often
seen laying eggs, and I think I was not deceived in my obser-
tion that she dropped a bunch of eggs into the open ditch while
balancing herself just a little way above the surface of the
water. I have, also, seen her settled upon the reeds in brackish
water with her abdomen submerged in part, and there attaching
a cluster of eggs. I feel pretty sure that L. auripennis does
not always deposit the whole of her eggs at one time, as I have
seen her attach a cluster of not more than a dozen small yellow

eggs. There must be more than one hundred eggs in one of
the large bunches. The eggs of some of the Agrions are bright
apple-green, but I cannot be sure that I have ever seen them in
the very act of oviposition. They have curious habits of settling upon leaves and grass growing in the water, and often
allow their abdomens to fall below the surface of the water;
sometimes they fly against the surface, but I never saw what I
could assert to be the project-
ing of the eggs from the body
upon plants or into the water.
The English entomologists
assert that the female Agrion
goes below the surface to a
depth of several inches to
deposit eggs upon the submerged stems of plants." The
Agrions, however, according
to Lacaze Duthiers, a French
anatomist, make, with the ovipositor, a little notch in the
plant upon which they lay their
eggs.

These eggs soon hatch, probably during the heat of summer. The larva is very active
in its habits, being provided
with six legs, attached to the
thorax, on the back of which
are the little wing-pads, or
rudimentary wings. The large
head is provided with enormous eyes, while a pair of simple, minute eyelets (ocelli) are

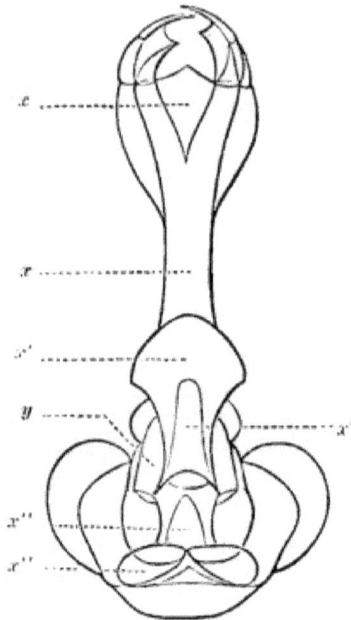

127. Under side of head of Diplax,
with the labium or mask fully extended. *x, x', x''*, the three subdi-
visions of the labium. *y*, the maxillæ or second pair of jaws.

placed near the origin of the small bristle-like feelers, or antennæ. Seen from beneath, instead of the formidable array of
jaws and accessory organs commonly observed in most carnivorous larvæ, we see nothing but a broad, smooth mask covering
the lower part of the face; as if from sheer modesty our young
Dragon fly was endeavoring to conceal a gape. But wait a
moment. Some unwary insect comes within striking distance.
The battery of jaws is unmasked, and opens upon the victim.

This mask (Fig. 127) is peculiar to the young, or larva and pupa of the Dragon fly. It is the labium, or under lip greatly enlarged, and armed at the broad spoon-shaped extremity (Fig. 127, x) with two sharp hooks, adapted for seizing and retaining its prey. At rest, the terminal half is so bent up as to conceal the face, and thus the creature crawls about, to all appearance, the most innocent and lamb-like of insects.

Not only does the immature Dragon fly walk over the bottom of the pool or stream it inhabits, but it can also leap for a considerable distance, and by a most curious contrivance. By a syringe-like apparatus lodged in the end of the body, it discharges a stream of water for a distance of two or three inches behind it, thus propelling the insect forwards. This apparatus combines the functions of locomotion and respiration. There are, as usual, two breathing pores (stigmata) on each side of the thorax. But the process of breathing seems to be mostly carried on in the tail. The trachea are here collected in a large mass, sending their branches into folds of membrane lining the end of the alimentary canal, and which act like a piston to force out the water. The entrance to the canal is protected by three to five triangular horny valves (Fig. 128, 9, 10, 128 a, side view), which open and shut at will. When open, the water flows in, bathing the internal gill-like organs, which extract the air from the water, which is then suddenly expelled by a strong muscular effort.

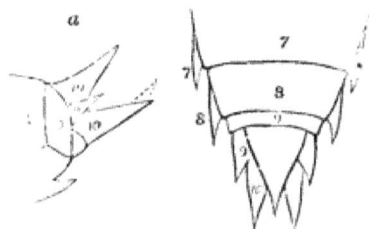

128. Abdominal valves; a, side view.

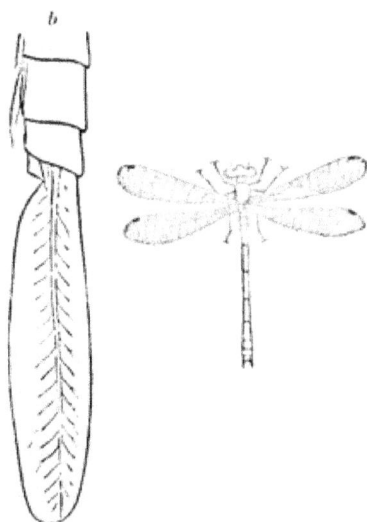

129. Agrion; b, False Gill of Larva.

In the smaller forms, such as Agrion (A. saucium, Fig. 129;

10

Fig. 129 $b$, side view of false gill, showing but one leaf), the respiratory leaves, called the tracheary, or false-gills, are not enclosed within the body, but form three broad leaves, permeated by tracheæ, or air-vessels. They are not true gills, however, as the blood is not aerated in them. They only absorb air to supply the tracheæ, which aerate the blood only within the general cavity of the body. These false gills also act as a rudder to aid the insect in swimming.

It is interesting to watch the Dragon flies through their transformations, as they can easily be kept in aquaria. Little, almost nothing, is known regarding their habits, and any one who can spend the necessary time and patience in rearing them, so as to trace up the different stages from the larva to the adult fly, and describe and figure them accurately, will do good service to science.

130. Pupa of Cordulia.

Mr. Uhler states that at present we know but little of the young stages of our species, but the larva and pupa of the Libellulas may be always known from the Æschnas by the shorter, deeper and more robust form, and generally by their thick clothing of hair. Figure 130 represents the pupa of Cordulia lateralis, and figure 131 that of a Dragon fly referred doubtfully to the genus Didymops.

131. Pupa of Didymops?

For descriptions and figures of other forms the reader may turn to Mr. Louis Cabot's essay "On the Immature State of the Odonata," published by the Museum of Comparative Zoology at Cambridge.

The pupa scarcely differs from the larva, except in having larger wing-pads (Fig. 132). It is still active, and as much of a gourmand as ever. When the insect is about to assume the pupa state, it moults its skin. The body having outgrown the larva skin, by a strong muscular effort a rent opens along the back of the thorax, and the insect having fastened its claws into some object at the bottom of the pool, the pupa gradually works its way out of the larva-skin. It is now considerably larger than before. Immediately after this tedious operation, its body is soft, but the crust soon hardens. This change, with most species, probably occurs early in summer.

When about to change into the adult fly, the pupa climbs up some plant near the surface of the water. Again its back yawns wide open, and from the rent our Dragon fly slowly emerges. For an hour or more, it remains torpid and listless, with its flabby, soft wings remaining motionless. The fluids leave the surface, the crust hardens and dries, rich

132. Pupa of Æschna.

and varied tints appear, and our Dragon fly rises into its new world of light and sunshine a gorgeous, but repulsive being. Tennyson thus describes these changes in "The Two Voices":—

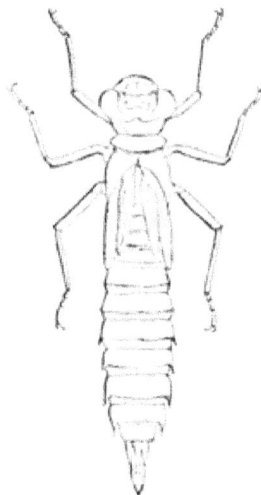

To-day I saw the Dragon fly
Come from the wells where he did lie.
An inner impulse rent the veil
Of his old husk: from head to tail
Came out clear plates of sapphire mail.

He dried his wings; like gauze they grew;
Through crofts and pastures wet with dew
A living flash of light he flew.

Of our more common, typical forms of Dragon flies, we figure a few, commonly observed during the summer. The three-spotted Dragon fly (Libellula trimaculata), of which figure 133 represents the male, is so called from the three dark clouds on the wings of the female. But the opposite sex differs in having a dark patch at the front edge of the wings, and a single broad cloud just beyond the middle of the wing.

Libellula quadrimaculata, the four-spotted Dragon fly (Fig.

134), is seen on the wing in June, flying through dry pine woods far from any standing water.

The largest of our Dragon flies are the "Devil's Darning-

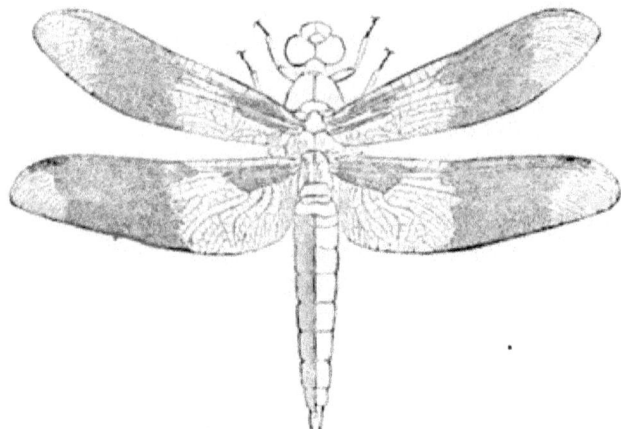

133. Libellula trimaculata, male.

needles," Eschna heros and grandis, seen hawking about our gardens till dusk. They frequently enter houses, carrying dis-

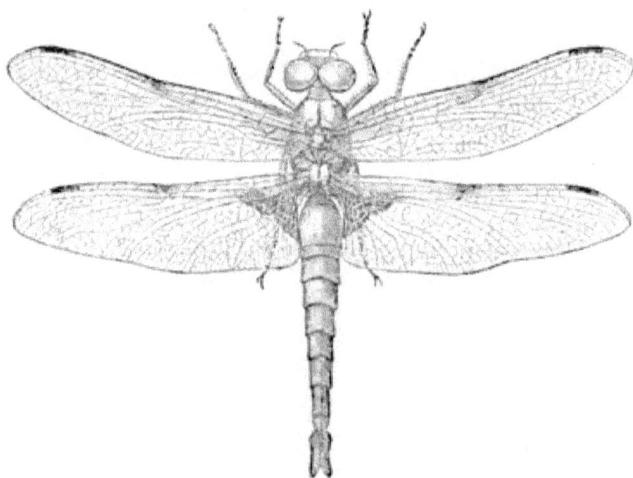

134. Libellula quadrimaculata.

may and terror among the children. The hind-body is long and cylindrical, and gaily colored with bright green and bluish bands and spots.

One of our most common Dragon flies is the ruby Dragon fly, Diplax rubicundula, which is yellowish-red. It is seen everywhere flying over pools, and also frequents dry sunny woods and glades. Another common form is Diplax Berenice (Fig. 135 male, Fig. 136 female. The accompanying cut (137) represents the larva, probably of this species, according to Mr. Uhler.) It is black, the head blue in front, spotted with yellow, while the thorax and abdomen are striped. w i t h  y e l l o w .

135. Diplax Berenice, male.

There are fewer stripes on the body of the male, which has only four large yellow spots on each side of the abdomen. Still another pretty species is Diplax Elisa (Fig. 138). It is black, with the head yellowish and with greenish-yellow spots on the sides of the thorax and base of the abdomen. There are three dusky spots on the front edge of each wing, and a large cloud at the base of the hind pair towards the hind angles of the wing.

137. Larva of Diplax.

Rather a rare form, and of much smaller stature is the Nannophya bella (Fig. 138, female). It was first detected in Baltimore, and we afterwards found it not unfrequently by a pond in Maine. Its abdomen is unusually short, and the reticulations o f  t h e wings are large and simple. The female is black. while the male is frosted over with a whitish powder. Many more species of this family are found in this country, and for descriptions of them we

136. Diplax Berenice, female.

would refer the reader to Dr. Hagen's "Synopsis of the Neuroptera of North America," published by the Smithsonian Institution.

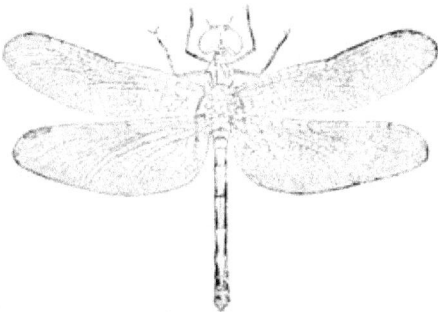

The Libellulidæ, or family of Dragon flies, and the Ephemer-
idæ, or May flies (Fig. 140), are the most characteristic of the
Neuroptera, or veiny-winged insects. This group is a most
interesting one to the
systematist, as it is
composed of so many
heterogeneous forms
which it is almost
impossible to classify
in our rigid and at
present necessarily
artificial systems.
We divide them into
families and sub-fam-
ilies, genera and sub-

138. Diplax Elisa.

genera, species and varieties, but there is an endless shifting
of characters in these groups. The different groups would seem
well limited after studying certain
forms, when to the systematist's sorrow, here comes a creature, perhaps
mimicking an ant, or aphis, or other
sort of bug, or even a butterfly, and
for which they would be readily mistaken by the uninitiated. Bibliogra-
phers have gone mad over books that
could not be classified. Imagine the despair of an insect-hunter

139. Nannophya bella.

and entomophile, as he sits down to
his box of dried neuroptera. He seeks
for a true neuropter in the white ant
before him, but its very form and
habits summon up a swarm of true
ants; and then the little wingless book
louse (Atropos, Fig. 141) scampering
irreverently over the musty pages of
his Systema Naturæ, reminds him
of that closest friend of man — Pediculus vestimenti. Again, his studies
lead him to that gorgeous inhabitant

140. May Fly.

of the South, the butterfly-like Ascalaphus, with its resplendent
wings, and slender, knobbed antennæ so much like those of
butterflies, and visions of these beautiful insects fill his mind's

eye; or sundry dun-colored caddis flies, modest, delicate neuroptera, with finely fringed wings and slender feelers, create
doubts as to whether they are not really allies of the clothes
moth, so close is the
resemblance.

Thus the student is
constantly led astray by
the wanton freaks Nature
plays, and becomes scep-
tical as regards the truth
of a natural system,
though there is one to
be discovered; and at
last disgusted with the
stiff and arbitrary sys-
tems of our books,— a
disgust we confess most
wholesome, if it only
leads him into a closer
communion with nature.
The sooner one leaves
those maternal apron-

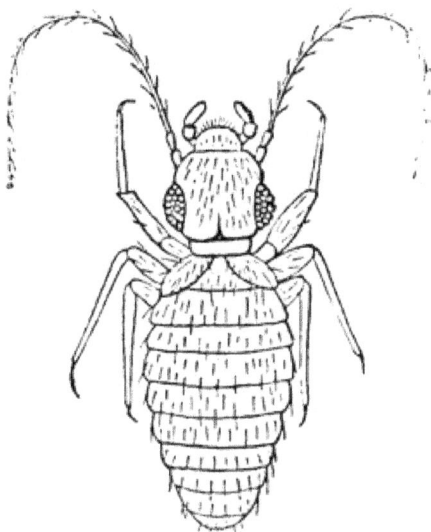

111. Death Tick.

strings,— books,— and learns to identify himself with nature,
and thus goes out of himself to affiliate with the spirit of the
scene or object before him, — or, in other words, cultivates
habits of the closest observation and most patient reflection,—
be he painter or poet, philosopher or insect-hunter of low
degree, he will gain an intellectual strength and power of inter-
preting nature, that is the gift of true genius.

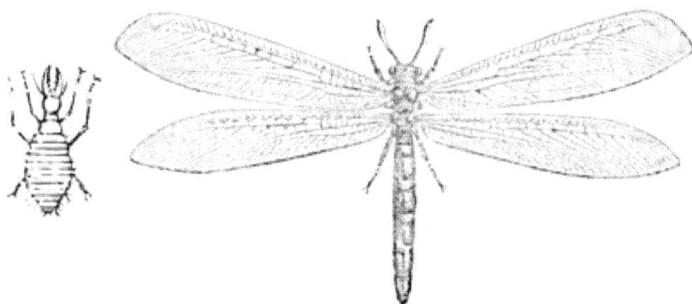

The Ant Lion and adult.

# CHAPTER XI.

But few naturalists have busied themselves with the study of mites. The honored names of Hermann, Von Heyden, Dugés, Dujardin and Pagenstecher, Nicolet, Koch and Robin, and the lamented Claparède of Geneva, lead the small number who have published papers in scientific journals. After these, and except an occasional note by an amateur microscopist who occasionally pauses from his "diatomaniacal" studies, and looks upon a mite simply as a "microscopic object," to be classed in his micrographic Vade Mecum with mounted specimens of sheep's wool, and the hairs of other quadrupeds, a distorted proboscis of a fly, and podura scales, we read but little of mites and their habits. But few readers of our natural history text-books learn from their pages any definite facts regarding the affinities of these humble creatures, their organization and the singular metamorphosis a few have been known to pass through. We shall only attempt in the present article to indicate a few of the typical forms of mites, and sketch, with too slight a knowledge to speak with much authority, an imperfect picture of their appearance and modes of living.

Mites are lowly organized Arachnids. This order of insects is divided into the Spiders, the Scorpions, the Harvestmen and the Mites (Acarina). They have a rounded oval body, without the usual division between the head-thorax and abdomen observable in spiders, the head-thorax and abdomen being merged in a single mass. There are four pairs of legs, and the mouth parts consist, as seen in the adjoining figure of a young tick (Fig. 142, young Ixodes albipictus), of a pair of maxillæ (c), which in the adult terminates in a two or three-jointed palpus, or feeler; a pair of mandibles (b), often covered with several rows of fine teeth, and ending in three or four larger

(116)

hooks and a serrated labium (*a*). These parts form a beak which the mite or tick insinuates into the flesh of its host, upon the blood of which it subsists. While many of the mites are parasitic on animals, some are known to devour the eggs of insects and other mites, thrusting their beaks into the egg, and sucking the contents. We have seen a mite (Nothrus ovivorus. Fig. 143) busily engaged in destroying the eggs of a moth like that of the Canker worm, and Dr. Shimer has observed the Acarus? malus sucking the eggs of the Chinch bug. I have also observed another mite devouring the Aphides on the rose leaves in my garden, so that a few mites may be set down as beneficial to vegetation. While a few species are injurious to man, the larger part are beneficial, being either parasitic, and baneful to other noxious animals, or more directly useful as scavengers, removing decaying animal and vegetable substances.

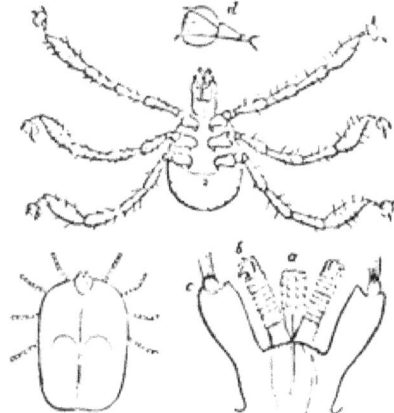
142. Ixodes albipictus and young.*

The transformations of the mites are interesting to the philosophic zoologist, since the young of certain forms are remarkably different from the adults, and in reaching the perfect state the mite passes through a metamorphosis more striking than that of many insects. The young on leaving the egg have six legs, as we have seen in the case of the Ixodes. Sometimes, however, as. for example, in the larva, as we may call it, of a European mite, Typhlodromus pyri, the adult of which, according to A. Scheuten, is allied to Acarus, and lives under the epidermis of the leaves of the pear in Europe (while Mr. T. Taylor, of the Department of Agriculture at Washington, has found a species in the pear leaves about Washington, and still another form in peach leaves), there are but two pairs of legs present, and the body is long, cylindrical and in a degree worm-like.

* The figure at the bottom on the left represents the adult, fully-gorged tick.

I have had the good fortune to observe the different stages
of a bird mite, intermediate in its form between the Acarus and
Sarcoptes, or Itch mite.   On March 6th, Mr. C. Cooke called my
attention to certain little mites which were situated on the narrow groove between the main stem of the barb and the outer
edge of the barbules of the feathers of the Downy Woodpecker, and subsequently we found the other forms in the down
under the feathers.   These long worm-like mites were evidently
the young of a singular Sarcoptes-like mite, as they were found
on the same specimen of Woodpecker at about the same date,
and it is known that the growth of mites is rapid, the metamorphoses, judging by the information which we now possess, occupying usually but a few days.

The young (though there is, probably, a still earlier hexapodous stage) of this Sarcoptid has an elongated, oblong, flattened

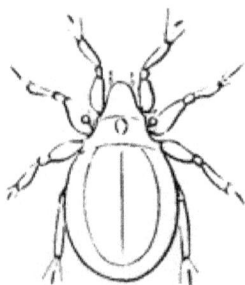

143. Egg-eating Mite.

body, with four short legs, provided with
a few bristle-like hairs, and ending in a
stalked sucker, by aid of which the mite
is enabled to walk over smooth, hard
surfaces.   The body is square at the
end, with a slight median indentation,
and four long bristles of equal length.
They remained motionless in the groove
on the barb of the feather, and when
removed seemed very inert and sluggish.
A succeeding stage of this mite, which
may be called the pupal, is considerably smaller than the larva
and looks somewhat like the adult, the body having become
shorter and broader.   The adult is a most singular form, its
body being rudely ovate, with the head sunken between the
fore legs, which are considerably smaller than the second pair,
while the third pair are twice as large as the second pair, and
directed backwards, and the fourth pair are very small, not
reaching the extremity of the body, which is deeply cleft and
supports four long bristles on each side of the cleft, while other
bristles are attached to the legs and body, giving the creature,
originally ill-shapen, a haggard, unkempt appearance.   The two
stigmata or breathing pores open near the cleft in the end of
the body, and the external opening of the oviduct is situated
between the largest and third pair of legs.   No males were
observed.   In a species of Acarus (Tyroglyphus), somewhat

like the Cheese mite, which we have alive at the time of writing, in a box containing the remains of a Lucanus larva, which they seem to have consumed, as both young and old are swarming there by myriads, the young are oval and like the adults, except that they are six-legged, the fourth pair growing out after a succeeding moult.

Such is a brief summary of what has been generally known regarding the metamorphoses of a few species of mites. In a few kinds no males have been found; the females have been isolated after being hatched, and yet have been known to lay eggs, which produced young without the interposition of the males. This parthenogenesis has been noticed in several species.

These insects often suddenly appear in vast numbers on various articles of food and about houses, so as to be very annoying. Mr. J. J. H. Gregory, of Marblehead, Mass., has found a mite allied to the European species here figured (Fig. 144) very injurious to the seeds of the cabbage, which

144. Cheyletus.

it sucked dry. This is an interesting form, and we have called it Cheyletus seminivorus. It is of medium size, and especially noticeable from the tripartite palpi, which are divided into an outer, long, curved, claw-like lobe, with two rounded teeth at the base, and two inner, slender lobes pectinated on the inner side, the third innermost lobe being minute. The beak terminates in a sharp blade-like point.

We have received a Cheyletus-like mite, said to have been "extracted from the human face" in New Orleans. The body is oblong, square behind; the head is long and pointed, while the maxillæ end in a long, curved, toothed, sickle-like blade. That this creature has the habits of the itch mite is suggested by the

curious, large, hair-like spines with which the body and legs are
sparsely armed, some being nearly half as long as the body.
These hairs are covered with very fine spinules. Those on the
end of the body are regularly spoon-shaped. These strange
hairs, which are thickest on the legs, probably assisted the mite
in anchoring itself in the skin of its host. We have read no
account of this strange and interesting form. It is allied to the
Acaropsis Mericourti which lines in the human face.

A species, "apparently of the genus Gamasus," according to
Dr. Leidy, has been found living in the ear (at the bottom of
the external auditory meatus, and attached to the membrana
tympani) of steers. "Whether this mite is a true parasite of
the ear of the living ox, or whether it obtained access to the
position in which it was found after the death of the ox in the
slaughter house, has not yet been determined."

We will now give a hasty glance at the different groups of
mites, pausing to note those most interesting from their habits
or relation to man.

The most highly organized mite (and by its structure most
closely allied to the spider) is the little red garden mite, belonging to the genus Trombidium, to which the genus Tetranychus
is also nearly related. Our own species of the former genus
have not been "worked up," or in other words identified and
described, so that whether the European T. holosericeum Linn.
is our species or not, we cannot tell. The larvæ of this and
similar species are known to live parasitically upon Harvestmen (Phalangium), often called Daddy-long-legs; and upon
Aphides, grasshoppers and other insects. Mr. Riley has made
known to us through the "American Naturalist" (and from his
account our information is taken), the habits of certain young
of the garden mite (Trombidium) which are excessively annoying in the Southwestern States. The first is the Leptus? Americanus (Fig. 145), or American Harvest mite. It is only known
as yet in the larval or Leptus state, when it is of the form indicated in the cut, and brick red in color. "This species is barely
visible with the naked eye, moves readily and is found more
frequently upon children than upon adults. It lives mostly on
the scalp and under the arm pits, but is frequently found on the
other parts of the body. It does not bury itself in the flesh,
but simply insinuates the anterior part of the body just under
the skin, thereby causing intense irritation, followed by a little

red pimple. As with our common ticks, the irritation lasts only
while the animal is securing itself, and its presence would afterwards scarcely be noticed but for the pimple which results."

The second species (Fig. 145 *b*, Leptus? irritans) is also only
known in the Leptus stage. It is evidently the larva of a distinct genus from the other form, having enormous maxillæ and
a broad body; it is also brick red. Mr. Riley says that "this is
the most troublesome and, perhaps, best known of the two,
causing intense irritation and swelling on all parts of the body,
but more especially on the legs and around the ankles. Woe
betide the person who, after bathing in the Mississippi any-
where in this latitude, is lured to some green dressing-spot of
weeds or grass! He may, for the time, consider himself fortu-

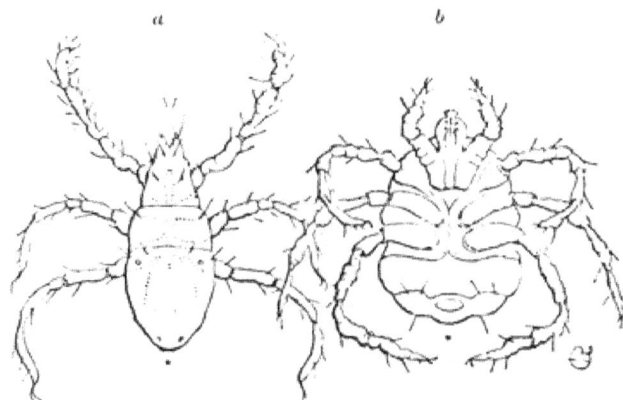

115 *a*. American Harvest Mite; *b*. Irritating Harvest Mite; the dots under-
neath indicating the natural size.

nate in getting rid of mud and dirt, but he will afterwards find to
his sorrow that he exchanged them for something far more tena-
cious in these microscopic Harvest-mites. If he has obtained
a good supply of them, he will in a few hours begin to suffer
from severe itching, and for the next two or three days will be
likely to scratch until his limbs are sore.

"With the strong mandibles and the elbowed maxillæ which
act like arms, this mite is able to bury itself completely in the
flesh, thereby causing a red swelling with a pale pustulous centre containing watery matter. If, in scratching, he is fortunate
enough to remove the mite before it enters, the part soon heals.
But otherwise the irritation lasts for two, three or four days,
the pustulous centre reappearing as often as it is broken.

11

"The animal itself, on account of its minute size, is seldom seen, and the uninitiated, when first troubled with it, are often alarmed at the symptoms and at a loss to account for them. Fortunately these little plagues never attach themselves to persons in such immense numbers as do sometimes young or so-called 'seed' ticks; but I have known cases where, from the irritation and consequent scratching, the flesh had the appearance of being covered with ulcers; and in some localities, where these pests most abound, sulphur is often sprinkled during 'jigger' season in the boots or shoes as a protection.

"Sulphur ointment is the best remedy against the effects of either of these mites, though when that cannot be obtained, saleratus water and salt water will partially allay the irritation.

"The normal food of either must, apparently, consist of the juices of plants, and the love of blood proves ruinous to those

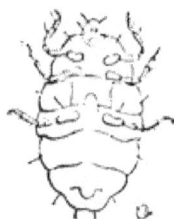

116. Astoma of the Fly.

individuals who get a chance to indulge it. For unlike the true Jigger, the female of which deposits eggs in the wound she makes, these Harvest-mites have no object of the kind, and when not killed by the hands of those they torment, they soon die victims to their sanguinary appetite."

Another Leptus-like form is the parasite of the fly, described by Mr. Riley under the name of Astoma? muscarum (Fig. 146). How nearly allied it is to the European Astoma parasiticum we have not the means of judging.

The European Tetranychus telarius Linn., or web-making mite, spins large webs on the leaves of the linden tree. Then succeed in the natural order the water mites (Hydrachna), which may be seen running over submerged sticks and on plants, mostly in fresh water, and rarely on the borders of the sea. The young after leaving the egg differ remarkably from the adults, so as to have been referred to a distinct genus (Achlysia) by the great French naturalist, Audouin. They live as parasites on various water insects, such as Dytiscus, Nepa and Hydrometra, and when mature live free in the water, though Von Baer observed an adult Hydrachna concharum living parasitically on the gills of the fresh-water mussel, Anodon. The species are of minute size. Collectors of beetles often meet with a species of Uropoda attached firmly to their specimens of dung-inhabiting or carrion beetles. It is a smoothly polished, round, flattened mite, with short, thick legs, scarcely reaching beyond the body.

We now come to the Ticks, which comprise the largest mites. The genus Argas closely resembles Ixodes. Gerstaecker states that the Argas Persicus is very annoying to travellers in Persia. The habits of the wood ticks (Ixodes) are well known. Travellers in the tropics speak of the intolerable torment occasioned by these pests which, occurring ordinarily on shrubs and trees, attach themselves to all sorts of reptiles, beasts and cattle, and even man himself as he passes by within their reach. Sometimes cases fall within the practice of the physician, who is called to remove the tick, which is found sometimes literally buried beneath the skin. Mr. J. Stauffer writes me, that "on June 23d the daughter of Abraham Jackson (colored), playing among the leaves in a wood, near Springville, Lancaster County, Penn., on her return home complained of pain in the arm. No attention was paid to it till the next day, when a raised tumor was noticed, a small portion protruding through the skin, apparently like a splinter of wood. The child was taken to Dr. Moreney, who applied the forceps, and after considerable pain

147. Cattle Tick.

to the child, and labor to himself, extracted a species of Ixodes, nearly one-quarter of an inch long, and of an oval form and brown mahogany color, with a metallic spot, like silver bronze, centrally on the dorsal region." This tick proved, from Mr. Stauffer's figures, to be, without doubt, Ixodes unipunctata. It has also been found in Massachusetts by Mr. F. G. Sanborn.

Another species is the Ixodes bovis (Fig. 147), the common cattle tick of the Western States and Central America. It is very annoying to horned cattle, gorging itself with their blood, but is by no means confined to them alone, as it lives indifferently upon the rattlesnake, the iguana, small mammals and undoubtedly any other animal that brushes by its lurking-place in the forest. It is a reddish, coriaceous, flattened, seed-like creature, with the body oblong oval, and contracted just behind the middle. When fully grown it measures from a quarter to half an inch in length. We have received it from Missouri, at the hands of Mr. Riley, and Mr. J. A. McNiel has found it very abundantly on horned cattle on the western coast of Nicaragua.

We now come to the genus Acarus (Tyroglyphus), of which the cheese and sugar mites are examples. Some species of Acarian mites have been found in the lungs and blood-vessels, and even the intestinal canal of certain vertebrates, while the too familiar itch insect lurks under the skin of the hand and other parts of the body of certain uncleanly human bipeds.

Many people have been startled by statements in newspapers and more authoritative sources, as to the immense numbers of mites (Acarus sacchari, Fig. 148) found in unrefined or raw sugar. According to Prof. Cameron, of Dublin, as quoted in the "Journal of the Franklin Institute," for November, 1868, "Dr. Hassel (who was the first to notice their general occurrence in the raw sugar sold at London) found them in a living state in no fewer than sixty-nine out of seventy-two samples. He did not detect them in a single specimen of refined sugar. In an inferior sample of raw sugar, examined in Dublin by Mr. Cameron, he reports finding five hundred mites in ten grains

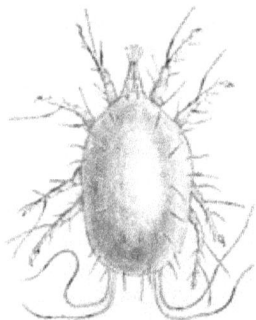

148. Sugar Mite.

of sugar, so that in a pound's weight occurred one hundred thousand of these little creatures, which seem to have devoted themselves with a martyr-like zeal to the adulteration of sugar. They appear as white specks in the sugar. The disease known as grocer's itch is, undoubtedly, due to the presence of this mite, which, like its ally the Sarcoptes, works its way under the skin of the hand, in this case, however, of cleanly persons. Mr. Cameron states that "the kind of sugar which is both health-ful and economical, is the dry, large-grained and light-colored variety."

Closely allied to the preceding, is the Cheese mite (Acarus siro Linn.), which often abounds in newly made cheese. Lyonet states that during summer this mite is viviparous. Acarus farinæ DeGeer, as its name indicates, is found in flour. Other species have been known to occur in ulcers.

We should also mention the Mange insect of the horse (Psoroptes equi, Fig. 149, much enlarged; *a*, head more magnified). According to Prof. Verrill it is readily visible to the naked eye and swarms on horses afflicted with the mange, which

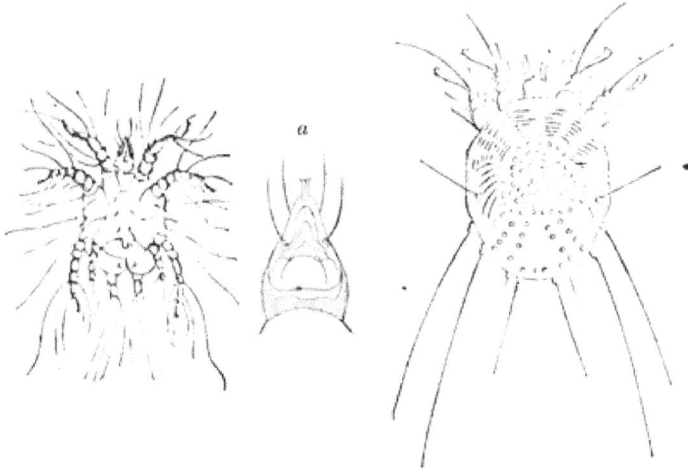

*a*

149. Mange Mite.                        150. Itch Mite.

is a disease analogous to the itch in man. It has a soft, depressed body, spiny beneath at the base of the legs and on the thorax. One or both of the two posterior pairs of feet bear suckers, and all are more or less covered with long, slender hairs. This insect may be destroyed by the same remedies as are used for lice and for the human itch. The best remedy is probably a solution of sulphuret of potassium.

The itch insect (Sarcoptes scabiei, Fig. 150) was first recognized by an Arabian author of the twelfth century, as the cause of the disease which results from its attacks. The body of the insect is rounded, with the two hind pair of feet rudimentary and bearing long hairs. It buries itself in the skin on the more protected parts of the body, and by its punctures maintains a constant irritation. Other species are known to infest the sheep and dog. Another singular mite is the Demodex follicnlorum (Fig. 151), which was discovered by Dr. Simon, of Berlin, buried in the diseased follicles of the wings of the nose in man. It is a long, slender, worm-like form, with

151. Nose Mite.

eight short legs, and in the larva state has six legs. This singular form is one of the lowest and most degraded of the order
of Arachnids.    A most singular mite was discovered by Newport on the body of a larva of a wild bee, and described by him
under the name of Heteropus ventricosus.    The body of the
fully formed female is long and slender.    After attaining this
form, its small abdomen begins to enlarge until it assumes a
globular form, and the mass of mites look like little beads.    Mr.
Newport was unable to discover the male, and thought that this
mite was parthenogenous.    It will be seen that the adult Demodex retains the elongated, worm-like appearance of the larva
of the higher mites, such as Typhlodromus.    This is an indication of its low rank, and hints of a relationship to the Tardigrades and the Pentastoma, the latter being a degraded mite,
and the lowest of its order, living parasitically within the bodies
of other animals.

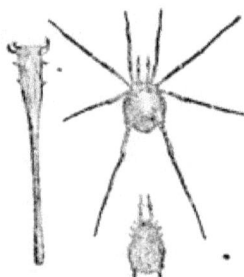

Harvestman.

# CHAPTER XII.

## BRISTLE-TAILS AND SPRING-TAILS.

The Thysanura, as the Poduras and their allies, the Lepismas, are called, have been generally neglected by entomologists, and but few naturalists have paid special attention to them.* Of all those microscopists who have examined Podura scales as test objects, we wonder how many really know what a Podura is?

In preparing the following account I have been under constant indebtedness to the admirable and exhaustive papers of Sir John Lubbock, in the London "Linnæan Transactions" (vols. 23, 26 and 27). Entomologists will be glad to learn that he is shortly going to press with a volume on the Poduras, which, in distinction from the Lepismas, to which he restricts the term Thysanura, he calls Collembola, in allusion to the sucker-like tubercle situated on the under side of the body, which no other insects are known to possess.

The group of Bristle-tails, as we would dub the Lepismas in distinction from the Spring-tails, we will first consider. They are abundant in the Middle States under stones and leaves in forests, and northward are common in damp houses, while one

---

* Nicolet, in the "Annales de la Société Entomologique de France" (tome v, 1847), has given us the most comprehensive essay on the group, though Latreille had previously published an important essay, "De l'Organization Extérieure des Thysanoures" in the "Nouvelles Annales du Museum d'Histoire Naturelle, Paris, 1832," which I have not seen. Gervais has also given a useful account of them in the third volume of "Aptères" of Roret's Suite a Buffon, published in 1844.

The Abbe Bourlet, Templeton, Westwood, and Haliday have published important papers on the Thysanura; and Meinert, a Danish naturalist, and Olfers, a German anatomist, have published important papers on the anatomy of the group. In this country Say and Fitch have described less than a dozen species, and the writer has described two American species of Campodea, C. Americana, our common form, and C. Cookei, discovered by Mr. C. Cooke in Mammoth Cave, while Humbert has described in a French scientific journal a species of Japyx (J. Saussurii) from Mexico.

beautiful species that we have never noticed elsewhere, is our "cricket on the hearth," abounding in the chinks and crannies of the range of our house, and also in closets, where it feeds on sugar, etc., and comes out like cockroaches, at night, shunning the light. Like the cockroaches, which it vaguely resembles in form, this species loves hot and dry localities, in distinction from the others which seek moisture as well as darkness. By some they are called "silver witches," and as they dart off, when disturbed, like a streak of light, their bodies being coated in a suit of shining mail, which the arrangement of the scales resembles, they have really a weird and ghostly look.

The most complicated genus, and the one which stands at the head of the family, is Machilis, one species of which lives in the Northern and Middle States, and another in Oregon. They affect damp places, living under leaves and stones. They all have rounded, highly arched bodies, and large compound eyes, the two being united together. The maxillary palpi are greatly developed, but the chief characteristics are the two-jointed stylets arranged in nine pairs along each side of the abdomen, reminding us of the abdominal legs of Myriopods. The body ends in three long bristles, as in Lepisma.

The Lepisma saccharina of Linnæus, if, as is probable, that is the name of our common species, is not uncommon in old damp houses, where it has the habits of the cockroach, eating cloths, tapestry, silken trimmings of furniture, and doing occasional damage to libraries by devouring the paste, and eating holes in the leaves and covers of books.

In general form Lepisma may be compared to the larva of Perla, a net-veined Neuropterous insect, and also to the narrow-bodied species of cockroaches, minus the wings. The body is long and narrow, covered with rather coarse scales, and ends in three many jointed anal stylets, or bristles, which closely resemble the many jointed antennæ, which are remarkably long and slender. The thermophilous species already alluded to may be described as perhaps the type of the genus, the L. saccharina being simpler in its structure. The body is narrow and flattened; the basal joints of the legs being broad, flat and almost triangular, like the same joints in the cockroaches. The legs consist of six joints, the tarsal joints being large and two in number, and bearing a pair of terminal curved claws. The three thoracic segments are of nearly equal size, and the eight abdominal segments are also of similar size. The tracheæ are well developed, and may be readily seen in the legs. The end of the rather long and weak abdomen is propped up by two or three pairs of bristles, which are simple, not jointed, but moving freely at their insertion; thus they take the place of legs, and remind one of the abdominal legs of the Myriopods; and we shall see in certain other genera (Machilis and Campodea) of the Bristle-tails that there are, actually two-jointed bristles arranged in pairs along the abdomen. They may probably be directly compared with the abdominal legs of Myriopods. Further study, however, of the homologies of these peculiar appendages, and especially a knowledge of the embryological development of Lepisma and Machilis, is needed before this interesting point can be definitely settled. The three many jointed anal stylets may, however, be directly compared with the similar appendages of Perla and Ephemera. The mode of insertion of the antennæ of this family is much like that of the Myriopods, the front of the head being flattened, and concealing the base of the antennæ, as in the Centipedes and Pauropus. Indeed, the head of any Thysanurous insect seen from above, bears a general resemblance in some of its features to that of the Centipede and its allies. So in a less degree does the head of the larvæ of certain Neuroptera and Coleoptera. The eyes are compound, the single facets forming a sort of heap. The clypeus and labrum, or upper lip, is, in all the Thysanura, carried far down on the under side of the head, the clypeus being almost obsolete in the Poduridæ, this being one of the most essential characters of that family. Indeed, it is somewhat singular that these and other important characteristics of this group have been almost entirely passed over by authors, who have consequently separated these insects from other groups on what appear to the writer as comparatively slight and inconsiderable characters. The mouth-parts of the Lepismatidæ (especially the thermophilous Lepisma, which we now describe) are most readily compared with those of the larva of Perla. The rather large, stout mandibles are concealed at their tips, under the upper lip, which moves freely up and down when the creature opens its mouth. The mandible is about one-third as broad as long, armed with three sharp teeth on the outer edge, and with a broad cutting edge within, and still further inwards a lot

of straggling spinules. In all these particulars, the mandible of
Lepisma is comparable with that of certain Coleoptera and Neuroptera. So also are the maxillæ and labium, though we are
not aware that any one has indicated how close the homology
is. The accompanying figure (152) of the maxilla of a beetle
may serve as an example of the maxilla of the Coleoptera,
Orthoptera and Neuroptera. In these insects it consists almost
invariably of three lobes, the outer being the palpus, the middle
lobe the galea, and the innermost the lacinia; the latter under-
going the greatest modifications, forming a comb composed of
spines and hairs varying greatly in relative size and length.
How much the palpi vary in these groups of insects is well
known. The galea sometimes forms a palpus-like appendage.

Now these three lobes may be easily distinguished
in the maxilla of Lepisma. The palpus instead of
being directed forward, as in the insects mentioned
above (in the pupa of Ephemera the maxilla is
much like that of Lepisma), is inserted nearer the
base than usual and thrown off at right angles to
the maxilla, so that it is stretched out like a leg,
and in moving about the insect uses its maxillæ
partly as supports for its head. They are very long and large,
and five or six-jointed. The galea, or middle division, forms a
simple lobe, while the lacinia has two large chitinous teeth on
the inner edge, and internally four or five hairs arising from
a thin edge.

152. Maxilla.

The labium is much as in that of Perla, being broad and short,
with a distinct median suture, indicating its former separation
in embryonic life into a pair of appendages. The labial palpi
are three-jointed, the joints being broad, and in life directed
backwards instead of forwards as in the higher insects.

There are five American species of the genus Lepisma in the
Museum of the Peabody Academy. Besides the common L. saccharina? there are four undescribed species; one found about
outhouses and cellars, and the heat-loving form, perhaps an
imported species, found in a kitchen in Salem, and apparently
allied to the L. thermophila Lucas, of houses in Brest, France;
and lastly two allied forms, one from Key West, and another
from Polvon, Western Nicaragua, collected by Mr. McNiel.
The last three species are beautifully ornamented with finely
spinulated hairs arranged in tufts on the head; while the sides

of the body, and edges of the basal joints of the legs are fringed with them.

The interesting genus Nicoletia stands at the bottom of the group. It has the long, linear, scaleless body of Campodea, in the family below, but the head and its appendages are like Lepisma, the maxillary palpi being five-jointed, and the labial palpi four-jointed. The eyes are simple, arranged in a row of

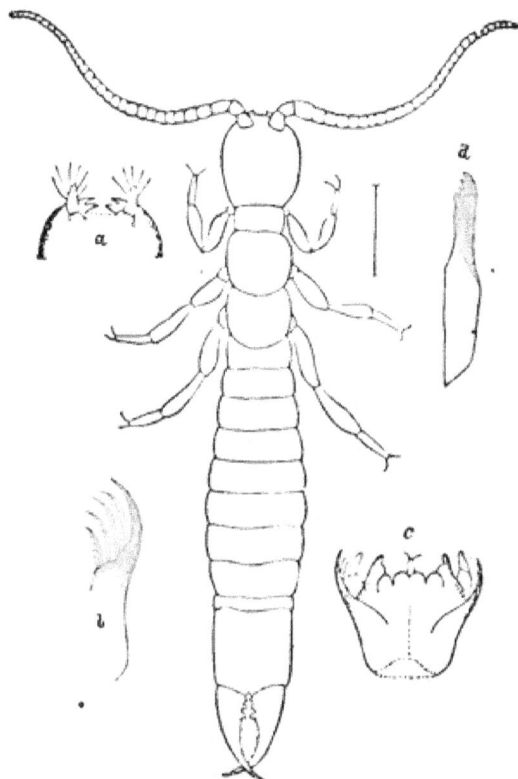

153. Japyx solifugus.

seven on each side of the head. The abdomen ends in three long and many jointed stylets, and there are the usual "false branchial feet" along each side of the abdomen. There are two European species which occur in greenhouses. No species have yet been found in America.

The next family of Thysanura is the Campodeæ, comprising the two genera Campodea and Japyx. These insects are much

smaller than the Lepismidæ, and in some respects are intermediate between that family and the Poduridæ (including the
Smynthuridæ).

In this family the body is long and slender, and the segments
much alike in size. There is a pair of spiracles on each thoracic
ring. The mandibles are long and slender, ending in three or

four teeth, and with the other appendages of the mouth are concealed
within the head, "only the tips of the
palpi (and of the maxillæ when these
are opened) projecting a very little
beyond the rounded entire margin of
the epistoma," according to Haliday.
The maxillæ are comb-shaped, due to
the four slender, minutely ciliated
spines placed within the outer tooth.
The labium in Japyx is four-lobed and
bears a small two-jointed palpus. The
legs are five-jointed, the tarsi consisting of a single joint, ending in two
large claws. The abdomen consists
of ten segments, and in Campodea
along each side is a series of minute,
two-jointed appendages such as have
been described in Machilis. These are
wanting in Japyx. None of the species in this family have the body cov

154. Campodea staphylinus.

ered with scales. They are white, with a yellowish tinge.

The more complicated genus of the two is Japyx (Fig. 153,
Japyx solifugus, found under stones in Southern Europe; a, the
mouth from beneath, with the maxillæ open; b, maxilla; d, mandible; e, outline of front of head seen from beneath, with the
labial palpi in position) which, as remarked by the late Mr. Haliday (who has published an elaborate essay on this genus in the
Linnæan Transactions, vol. 24, 1864), resembles Forficula in the
large forceps attached to its tail. An American species (J. Saussurii) lives in Mexico, and we look for its discovery in Texas.

Campodea (C. staphylinus Westw., Fig. 154, enlarged; a,
mandible; b, maxilla), otherwise closely related, has more rudimentary mouth-parts, and the abdomen ends in two many
jointed appendages.

Our common American species of Campodea (C. Americana) lives under stones in damp places. It is yellowish, about a sixth of an inch in length, is very agile in its movements, and would easily be mistaken for a very young Lithobius. A larger species and differing in having longer antennæ, has been found by Mr. C. Cooke in Mammoth Cave, and has been described in the "American Naturalist" under the name of Campodea Cookei. Haliday has remarked that this family bears much resemblance to the Neuropterous larva of Perla (Fig. 155), as previously remarked by Gervais; and the many points of resemblance of this family and the Lepismidæ to the larval forms of some Neuroptera that are active in the pupa state (the Pseudoneuroptera of Erichson and other authors) are very striking. Campodea resembles the earliest larval form of Chloëon, as figured by Sir John Lubbock, even to the single jointed tarsus; and why these two Thysanurous families should be removed from the Neuroptera we are unable, at present, to understand, as to our mind they scarcely diverge from the Neuropterous type more than the Mallophaga, or biting lice, from the type of Hemiptera.

Haliday, remarking on the opinion of Linnæus and Schrank, who referred Campodea to the old genus Podura, says

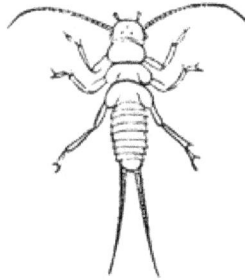

Fig. 155. Larva of Perla.

with much truth, "it may be perhaps no unfair inference to draw, that the insect in question is in some measure intermediate between both," i. e., Podura and Lepisma. This is seen especially in the mouth-parts which are withdrawn into the head, and become very rudimentary, affording a gradual passage into the mouth-parts of the Poduridæ, which we now describe.

The next group, the Podurelles of Nicolet, and Collembola of Lubbock, are considered by the latter, who has studied them with far more care than any one else, as "less closely allied" to the Lepismidæ "than has hitherto been supposed." He says "the presence of tracheæ, the structure of the mouth and the abdominal appendage, all indicate a wide distinction between the Lepismidæ and the Poduridæ. We must, indeed, in my opinion, separate them entirely from one another; and I would

12

venture to propose for the group comprised in the old genus
Podura, the term Collembola, as indicating the existence of a
projection, or mammilla, enabling the creature to attach or glue
itself to the body on which it stands." Then without expressing
his views as to the position and affinities of the Lepismidæ, he
remarks "as the upshot of all this, then, while the Collembola
are clearly more nearly allied to the Insecta than to the Crustacea or Arachnida, we cannot, I think, regard them as Orthoptera or Neuroptera, or even as true insects. That is to say, the
Coleoptera, Orthoptera, Neuroptera, Lepidoptera, etc., are in my
opinion, more nearly allied to one another than they are to the
Poduridæ or Smynthuridæ. On the other hand, we certainly
cannot regard the Collembola as a group equivalent in value to
the Insecta. If, then, we attempt to map out the Articulata, we
must, I think, regard the Crustacea and Insecta as continents,
the Myriopoda and Collembola as islands — of less importance,
but still detached. Or, if we represent the divisions of the
Articulata like the branching of a tree, we must picture the Collembola as a separate branch, though a small one, and much
more closely connected with the Insecta than with the Crustacea
or the Arachnida." Lamarck regarded them as more nearly
allied to the Crustacea than Insecta. Gervais, also, in the "Histoire Naturelle des Insectes: Aptères," indicates a considerable
diversity existing between the Lepismidæ and Poduridæ, though
they are placed next to each other. Somewhat similar views
have been expressed by so high an authority as Professor Dana,
who, in the "American Journal of Science" (vol. 37, Jan., 1864),
proposed a classification of insects based on the principle of
cephalization, and divided the Hexapodous insects into three
groups: the first (Ptero-prosthenics, or Ctenopters) comprising
the Hymenoptera, Diptera, Aphaniptera (fleas), Lepidoptera,
Homoptera, Trichoptera and Neuroptera; the second group
(Ptero-metasthenics, or Elytropters) comprising the Coleoptera,
Hemiptera and Orthoptera; while the Thysanura compose the
third group. Lubbock has given us a convenient historical
view of the opinions of different authors regarding the classification of these insects, which we find useful. Nicolet, the naturalist who, previous to Lubbock, has given us the most correct
and complete account of the Thysanura, regarded them as an
order, equivalent to the Coleoptera or Diptera, for example. In
this he followed Latreille, who established the order in 1796.

The Abbé Bourlet adopted the same view. On the other hand Burmeister placed the Thysanura as a separate tribe between the Mallophaga (Bird Lice) and Orthoptera, and Gerstaecker placed them among the Orthoptera. Fabricius and Blainville put them with the Neuroptera, and the writer, in his "Guide to the Study of Insects," and previously in 1863, ignorant of the views of the two last named authors, considered the Thysanura as degraded Neuroptera, and noticed their resemblance to the larvæ of Perla, Ephemera, and other Neuroptera, such as Rhaphidia and Panorpa, regarding them as standing "in the same relation to the rest of the Neuroptera [in the Linnæan sense], as the flea does to the rest of the Diptera, or the lice and Thrips to the higher Hemiptera."

After having studied the Thysanura enough to recognize the great difficulty of deciding as to their affinities and rank, the writer does not feel prepared to go so far as Dana and Lubbock, for reasons that will be suggested in the following brief account of the more general points in their structure, reserving for another occasion a final expression of his views as to their classification.

The Poduridæ, so well known by name, as affording the scales used by microscopists as test objects, are common under stones and wet chips, or in damp places, cellars, mushrooms and about manure heaps. They need moisture, and consequently shade. They abound most in

156. Smynthurus.

spring and autumn, laying their eggs at both seasons, though most commonly in the spring. During a mild December, they may be found in abundance under sticks and stones, even in situations so far north as Salem, Mass.

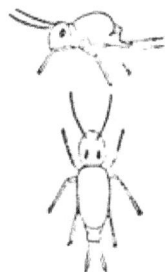

The body of the Poduras is rather short and thick, most so in Smynthurus (Fig. 156), and becoming long and slender in Tomocerus and Isotoma. The segments are inclined to be of unequal size, the prothoracic ring sometimes becoming almost obsolete, and some of the abdominal rings are much smaller than others; while in Lipura and Anura, the lowest forms of the group, the segments are all much alike in size.

The head is in form much like that of certain larvæ of Neuroptera and of Forficula, an Orthopterous insect. The basal half of the head is marked off from the eye-bearing piece (epicranium) by a V-shaped suture* (Fig. 157, head of Degeeria; compare also the head of the larva of Forficula, Fig. 158, in which the suture is the same), and the insertion of the antennæ is removed far down the front, near the mouth, the clypeus being very short; this piece, so large and prominent in the higher insects, is not distinctly separated by suture from the surrounding parts of the head, thus affording one of the best distinctive characters of the Poduridæ.  The eyes are situated on top of the head just behind the antennæ, and are simple, consisting of a group of from five to eight or ten united into a mass in Smynthurus,

157. Head of De-
geeria.

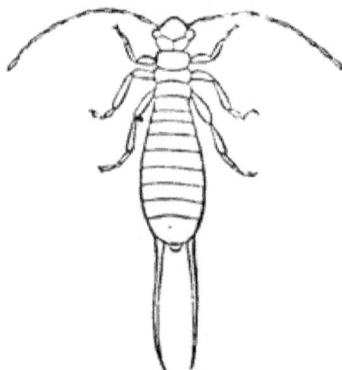

but separated in the Poduridæ (see Fig. 176, e, eye of Anurida). The antennæ are usually four-jointed, and vary in length in the different genera.

The mouth-parts are very difficult to make out, but by soaking the insect in potash for twenty-four hours, thus rendering the body transparent, they can be satisfactorily observed.  They are constructed on the same general type as the mouth-parts of the Neuroptera, Orthoptera and Coleoptera, and except in being degraded, and with certain parts

158. Larva of Forficula.

obsolete, they do not essentially differ.† On observing the living Podura, the mouth seems a simple ring, with a minute labrum and groups of hairs and spinules, which the observer, partly by

---

*The direct homology of these parts of the head (the occiput and the epicranium) with Perla, Forficula, etc., seems to me the best evidence we could have that the Poduræ are not an independent group.  In these most fundamental characters they differ widely from the Myriopods.  I am not aware that this important relation has been appreciated by observers.      •

†As we descend to the soft, tube-like, suctorial (?) mouth of Anura, which is said not to have hard mouth-parts, we see the final point of degradation to which the mouth of the Thysanura is carried.  I think that this gradual degradation of the mouth-parts in this group indicates that the appendages in these animals are not formed on an independent type, intermediate, so to speak, between the mandibulate and haustellate types, but are simply a modification (through disuse) of the mandibulate type as seen in Neuropterous insects.

guess-work, can identify as jaws and maxillæ, and labium.
But in studying the parts rendered transparent, we can identify
the different appendages. Figure 159 shows the common Tomocerus plumbeus greatly enlarged (Fig. 160, seen from above),
and as the mouth-parts of the whole group of Poduras are
remarkably constant, a description of one genus will suffice for
all. The labrum, or upper lip, is separated by a deep suture
from the clypeus, and is trapezoidal in form. The mandibles
and maxillæ are long and slender, and buried in the head, with
the tips capable of being extended out from the ring surrounding the mouth for a very short distance. The mandibles (md,
Fig. 159) are like those of the Neuroptera, Orthoptera and
Coleoptera in their general form, the tip ending in from three
to six teeth (three on one mandible and six on the other), while
below, is a rough, denticulated molar surface, where the food
seized by the terminal teeth is triturated and prepared to be
swallowed. Just behind the mandibles are the maxillæ, which
are trilobate at the end, as in the three orders of insects above
named. The outer lobe, or palpus, is a minute membranous
tubercle ending in a hair (Fig. 161, mp), while the middle lobe,
or galea, is nearly obsolete, though I think I have seen it in
Smynthurus, where it forms a lobe on the outside of the lacinia.
The lacinia, or inner lobe (Fig. 161, le; 162, the same enlarged),
in Tomocerus consists of two bundles of spinules, one broad
like a ruffle, and the other slender, pencil-like, ending in an inner
row of spines, like the spinules on the lacinia of the Japyx and
Campodea and, more remotely, the laciniæ of the three sub-
orders of insects above referred to. There is also a horny,
prominent, three-toothed portion (Fig. 161, g). These homologies have never been made before, so far as the writer is
aware, but they seem natural, and suggested by a careful examination and comparison with the above-mentioned mandibulate
insects.

The spring consists of a pair of three-jointed appendages,
with the basal joints soldered together early in embryonic life,
while the other two joints are free, forming a fork. It is longest
in Smynthurus and Degeeria, and shortest in Achorutes (Fig.
172, b), where it forms a simple, forked tubercle; and is obsolete
in Lipura and Anura, its place being indicated by an oval scar.
The third joint varies in form, being hairy, serrate and knife-
like in form, as in Tomocerus (Fig. 159, a), or minute, with a

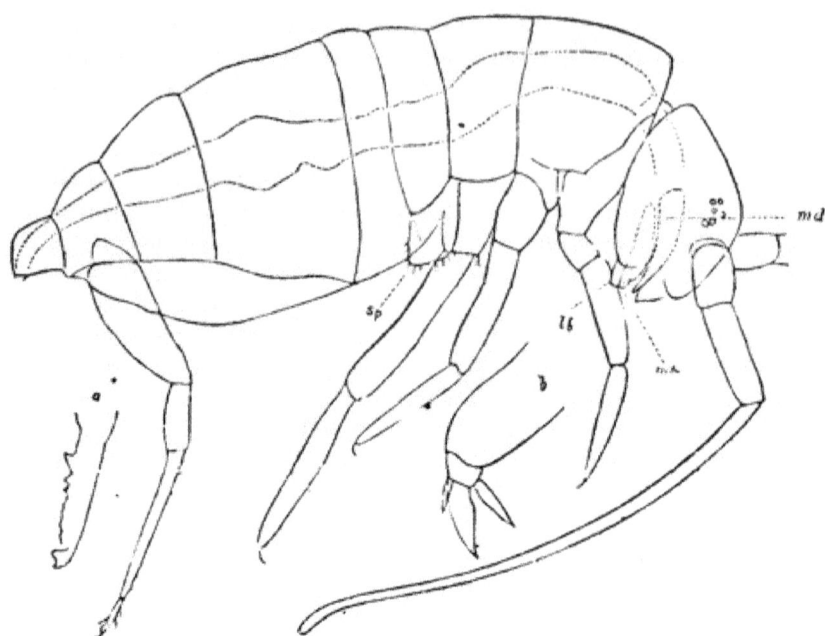

159.

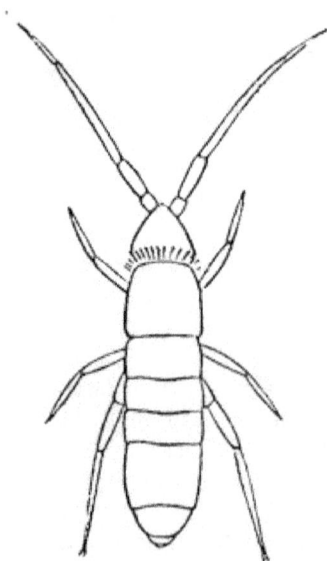

160.

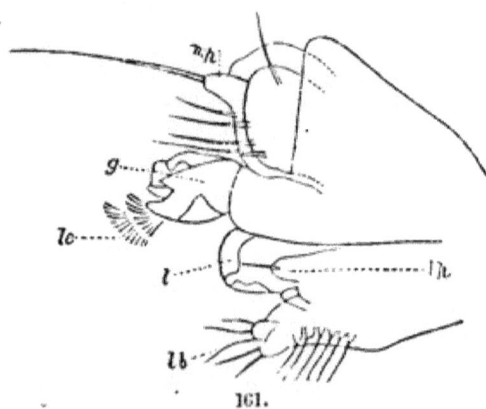

161.

162.

Tomocerus plumbeus and mouth-parts, greatly enlarged.

supplementary tooth, as in Achorutes (Fig. 172, *c*). This spring
is in part homologous with the ovipositor of the higher insects,
which originally consists of three pairs of tubercles, each pair
arising apparently from the seventh, eighth, and ninth (the latter the penultimate) segments of the abdomen in the Hymenoptera.    The spring of the Podura seems to be the homologue of
the third pair of these tubercles, and is inserted on the penultimate segment.   This comparison I have been able to make from
a study of the embryology of Isotoma.

Another organ, and one which, so far as I am aware, has been
overlooked by previous observers, I am disposed to consider as
possibly an ovipositor.  In the genus Achorutes, it may be found
in the segment just before the spring-bearing segment, and
situated on the median line of the body.  It consists (Fig. 163)
of two squarish valves, from between which
projects a pair of minute tubercles, or blades,
with four rounded teeth on the under side.
This pair of infinitesimal saws reminds one of
the blades of the saw-fly, and I am at a loss
what their use can be unless to cut and pierce
so as to scoop out a shallow place in which to
deposit an egg.   It is homologous in situation
with the middle pair of blades which composes
the ovipositor of higher insects, and if it should
prove to be used by the creature in laying its
eggs, we should then have, with the spring, an
additional point of resemblance to the Neurop-

163.  Catch holding
spring of Acho-
rutes.

tera and higher insects, and instead of this spring being an
important differential character, separating the Thysanura from
other insects, it binds them still closer, though still differing greatly in representing only a part of the ovipositor of the
higher insects.  (This is a catch for holding the spring in place.)

But all the Poduras differ from other insects in possessing a
remarkable organ situated on the basal segment of the abdomen.   It is a small tubercle, with chitinous walls, forming two
valves from between which is forced out a fleshy sucker, or, as
in Smynthurus, a pair of long tubes, which are capable of being
darted out on each side of the body, enabling the insect to
attach itself to smooth surfaces, and rest in an inverted position.

The eggs are laid few in number, either singly or several
together, on the under side of stones, chips or, as in the case

of Isotoma Walkerii, under the bark of trees. They are round,
transparent. The development of the embryo of Isotoma in
general accords with that of the Phryganeidæ and suggests on
embryological grounds the near relationship of the Thysanura
to the Neuroptera.

The earliest stage observed was at the time of the appearance
of the primitive band (Fig. 164, *a*, *b*, folding of the primitive

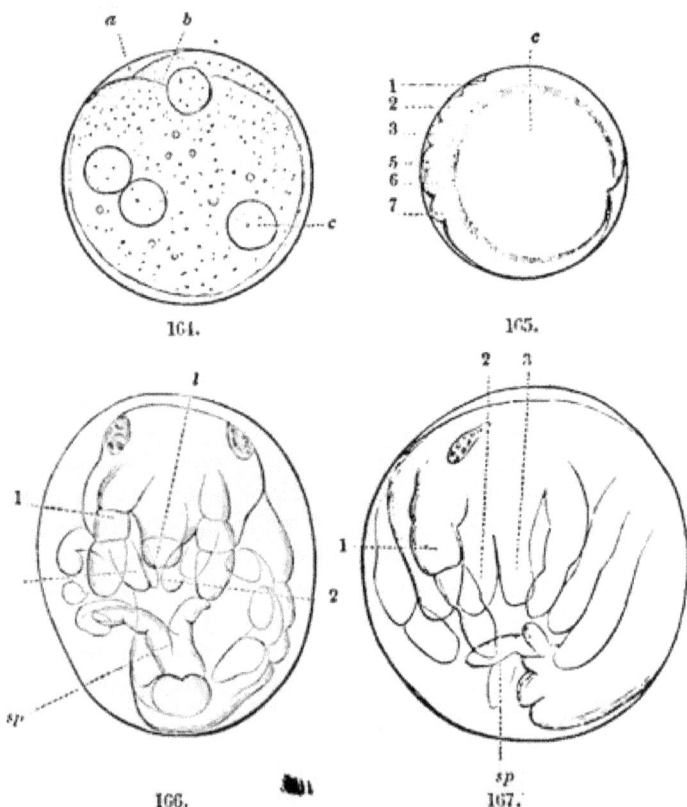

164.                    165.

166.                    167.

Development of a Poduran.

band; *c*, the dotted line crosses the primitive band, and terminates
in a large yolk granule) which surrounds the egg as in the Caddis
flies. Soon after, the primitive segments appear (Fig. 165; 1,
antennæ; 2, mandibles; 3, maxillæ; the labium was not seen; 5–7,
legs; *c*, yolk surrounded by the primitive band) and seem to originate just as in the Caddis flies. Figure 166 is a front view of

the embryo shortly before it is hatched; figure 167, side view of the same, the figures as in Fig. 165; *sp*, spring; *l*, labrum. The labrum or upper lip, and the clypeus are large and as distinct as in the embryos of other insects, a fact to which we shall allude again. The large three-jointed spring is now well developed, and the inference is drawn that it represents a pair of true abdominal legs. The embryo when about to hatch throws off the egg-shell and amnion in a few seconds. The larva is perfectly white and is very active in its movements, running over the damp, inner surface of the bark. It is a little over a hundredth of an inch in length, and differs from the adult in being shorter and thicker, with the spring very short and stout. In fact the larva assumes the form of the lower genera of the family, such as Achorutes and Lipura, the adult more closely resembling Degeeria. The larva after its first moult retains its early clumsy form, and is still white. After a second moult it becomes purplish, and much more slender, as in the adult. The eggs are laid and the young hatched apparently within a period of from six to ten days.

Returning to the stage indicated by figures 166 and 167, I am induced to quote some remarks published in the Memoirs of the Peabody Academy of Science, No. 2, p. 18, which seem to support the view that these insects are offshoots from the Neuroptera.

"The front of the head is so entirely different from what it is in the adult, that certain points demand our attention. It is evident that at this period the development of the insect has gone on in all important particulars much as in other insects, especially the Neuropterous Mystacides as described by Zaddach. The head is longer vertically than horizontally, the frontal, or clypeal region is broad, and greater in extent than the epicranio-occipital region. The antennæ are inserted high up on the head, next the ocelli, falling down over the clypeal region. The clypeus, however, is merged with the epicranium, and the usual suture between them does not appear distinctly in after life, though its place is seen in figure 167 to be indicated by a slight indentation. The labrum is distinctly defined by a well marked suture, and forms a squarish, knob-like protuberance, and in size is quite large compared to the clypeus. From this time begins the process of degradation, when the insect assumes its Thysanurous characters, which consist in an approach to the form of the Myriopodous head, the front, or clypeal region being reduced to a minimum, and the antennæ and eyes brought in closer proximity to the mouth than in any other insects."

Sir John Lubbock has given us an admirable account of the internal anatomy of these little creatures, his elaborate and patient dissections filling a great gap in our knowledge of their internal structure. The space at our disposal only permits us to speak briefly of the respiratory system. Lubbock found a simple system of tracheæ in Smynthurus which opens by "two spiracles in the head, opposite the insertion of the antennæ," i. e., on the back of the head. (Von Olfers says that they open on the prothorax.) Nicolet and Olfers claim to have found tracheæ in several lower genera (Orchesella, Tomocerus, and Achorutes and allied genera), but Lubbock was unable to detect them, and I may add that I have not yet been able after careful search to find them either in living specimens, or those rendered transparent by potash.

Having given a hasty sketch of the external aspect of the Poduras, I extract from Lubbock's work a synopsis of the families and genera for the convenience of the student, adding the names of known American species, or indications of undescribed native forms.

SMYNTHURIDÆ.—Body globular or ovoid; thorax and abdomen forming one mass; head vertical or inclined; antennæ of four or eight segments. Eyes eight on each side, on the top of the head. Legs long and slender. Saltatory appendage with a supplementary segment.

Smynthurus. Antennæ four-jointed, bent at the insertion of the fourth, which is nearly as long as the other three, and appears to consist of many small segments. No conspicuous dorsal tubercles. (In this country Fitch has described five species: S. arvalis, elegans, hortensis, Novæboracensis, and signifer. Figure 156 represents a species found in Maine.)

Dicyrtoma. Antennæ eight-jointed, five before, three after the bend. Two dorsal tubercles on the abdomen.

Papirius.* Antennæ four-jointed, without a well-marked

---

* Lubbock considers that Papirius should be placed in a distinct family from Smynthurus, because it wants tracheæ. Their presence or absence scarcely seems to us to be a family character, as they are wanting in the Poduridæ, and are not essential to the life of these animals, while in other respects Papirius seems to differ but slightly from Smynthurus.

elbow, and with a short terminal segment offering the appearance of being many-jointed.

Poduridæ.—This family comprises those species of the old genus Podura, in which the mouth has mandibles [also maxillæ and a labium], and the body is elongated, with a more or less developed saltatory appendage at the posterior extremity.

Orchesella. Segments of the body unequal in size, more or less thickly clothed with clubbed hairs. Antennæ long, six-jointed. Eyes six in number on each side, arranged in the form of an S. (One or two beautiful species live about Salem.)

Degeeria. Segments of the body unequal in size, more or less thickly clothed by clubbed hairs. Antennæ longer than the head and thorax, filiform, four-jointed. Eyes eight in number on each side of the head. (Two species, Degeeria decem-fasciata, Pl. 10, Figs. 2, 3, and D. purpurascens, Figs. 4, 5, are figured in the "Guide to the Study of Insects." Figure 168 represents a species found in Salem, Mass., closely allied to the European D. nivalis. Five species are already known in New England.)

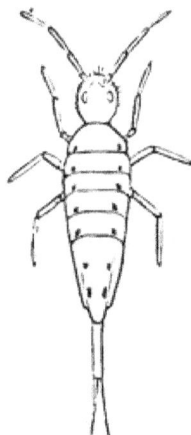

168. Degeeria.

Seira. Body covered with scales. Antennæ four-jointed; terminal segment not ringed. Eyes on a dark patch. Thorax not projecting over the head. Abdominal segments unequal.

Templetonia. Segments of the body subequal, clothed by clubbed hairs, and provided with scales. Antennæ longer than the head and thorax, five-jointed, with a small basal segment, and with the terminal portion ringed.

Isotoma. Four anterior abdominal segments subequal, two posterior ones small; body clothed with simple hairs and without scales. Antennæ four-jointed, longer than the head: segments subequal. Eyes seven in number on each side, arranged in the form of an S. (Three species are found in Massachusetts, one of which (I. plumbea) is figured on Pl. 10, Figs. 6, 7, of the "Guide to the Study of Insects," third edition.)

Tomocerus. Abdominal segments unequal, with simple hairs and scales. Antennæ very long, four-jointed, the two terminal segments ringed. Eyes seven in number on each side. (The European T. plumbea, Podura plumbea of authors, is our spe-

cies, and is common. Fig. 160, greatly enlarged, copied from
Templeton; Fig. 159, side view, see also Fig. 161, where the
mouth-parts are greatly enlarged, the lettering being the same,
*md*, mandibles; *mx*, maxillæ; *mp*, maxillary palpus; *lb*, labium;

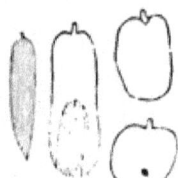

169. Scales of Tomocerus.                    171. Scale of Lepidocyrtus.

*lp*, labial palpus; *lc*, lacinia; *g*, portion ending in three teeth; *l*,
lobe of labium; *sp*, ventral sucking disk; the dotted lines passing
through the body represent the course of the intestine; *b*, end
of tibia, showing the tarsus, with the claw, and two accessory
spines; *a*, third joint of the spring.
Fig. 162, lacinia of maxilla greatly
enlarged. Fig. 169, different forms
of scales, showing the great variation in size and form, the narrow
ones running into a linear form,
becoming hairs. The markings
are also seen to vary, showing
their unreliable character as test
objects, unless a single scale is
kept for use.)

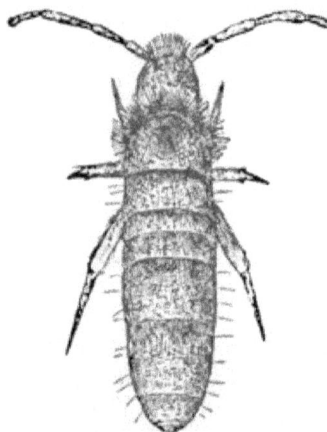

170. Lepidocyrtus.

Lepidocyrtus. Abdominal segment unequal, with simple hairs
and scales. Antennæ long, four-
jointed. Eyes eight in number
on each side. (Fig. 170, L. albi-
nos, an European species, from Hardwicke's "Science Gossip."
Fig. 171, a scale. Two species live in New England.)

Podura. Abdominal segments subequal. Hairs simple, no
scales. Antennæ four-jointed, shorter than the head. Eyes

eight in number on each side. Saltatory appendage of moderate
length.

Achorutes. Abdominal segments subequal. Antennæ short,
four-jointed. Eyes eight in number on each side. Saltatory
appendage quite short.

Figure 172 represents a species of this genus very abundant
under the bark of trees, etc., in New England. It is of a blackish
lead color; *a*, end of tibia bearing a tenant hair, with the tarsal
joint and large claw; *b*, spring; *c*, the third joint of the spring,
with the little spine at the base; figure 163, the supposed ovi-

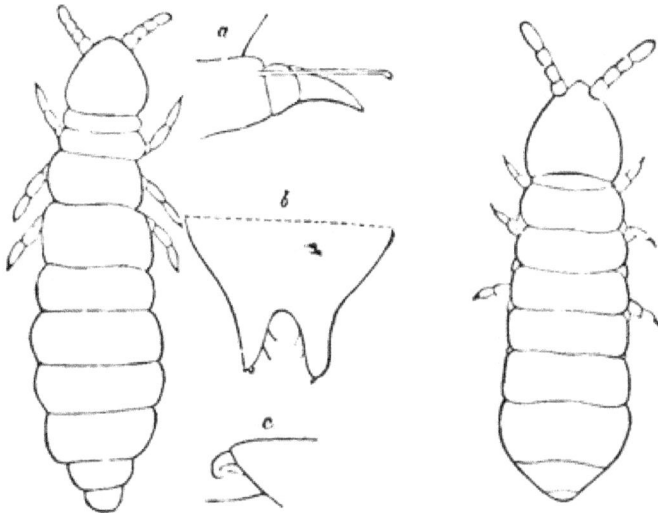

172. Achorutes.                173. Lipura fimetaria.

positor; *a*, the two blades spread apart; *b*, side view. The
mouth-parts in this genus are much as in Tomocerus, the maxillæ ending in a lacinia and palpus.

The three remaining genera, Lipura, Anurida and Anura, are
placed in the "family" Lipuridæ, which have no spring. Lubbock remarks that "this family contains as yet only two*
genera, Lipura (Burmeister), in which the. mouth is composed
of the same parts as those in the preceding genera, and Anura
(Gervais), in which the mandibles and maxillæ disappear." Our

---

*Dr. Laboulbene has recently, and we think with good reason, separated Anura
maritima from the genus Anura, under the name of Anurida maritima.

13

common white Lipura is the European L. fimetaria Linn. (Fig.
173, copied from Lubbock). The site of the spring is indicated
by an oval scar.

Figure 174 represents Anurida maritima found under stones
between tide marks at Nantucket. It is regarded the same as
the European species by Lubbock, to whom I had sent specimens
for comparison. This genus differs in the form of the head

Anurida maritima.

from Lipura and also wants the terminal upcurved spines, while
the antennæ are much more pointed. The legs (Fig. 175)
end in a large, long, curved claw. On examining specimens
soaked in potash, I have found that the mouth-parts of this
species (Fig. 176, *md*, mandibles; *mx*, maxillæ; *e*, eyes, and a
singular accessory group of small cells, are like those of Achorutes, as previously noticed by Laboulbène. The mandibles,
like those of other Poduras, end in from three to six teeth, and

have a broad, many-toothed molar surface below. The maxillæ
end in a tridentate lacinia as usual, though the palpi and galea
I have not yet studied.

The genus Anura may be readily recognized by the mouth
ending in an acutely conical beak, with its end quite free from
the head and hanging down beneath it. The body is short and
broad, much tuberculated, while the antennæ are short and
pointed, and the legs are much shorter than in Lipura, not
reaching more than a third of their length beyond the body.
Our common form occurs under the bark of trees.

For the reason that I can find no valid characters for separating these three genera as a family from the other Poduras, I
am inclined to think that they form, by the absence of the
spring, only a subdivision (perhaps a subfamily) of the Poduridæ.

The best way to collect Poduras is, on turning up the stick or
stone on the under side of which they live, to place a vial over
them, allowing them to leap into it; they may be incited to leap
by pushing a needle under the vial. They may also be collected by a bottle with a sponge saturated with ether or chloro-
form. They may be kept alive for weeks by keeping moist
slips of blotting paper in the vial. In this way I have kept
specimens of Degeeria, Tomocerus and Orchesella, from the
middle of December till late in January. During this time
they occasionally moulted, and Tomocerus plumbeus, after shed-
ding its skin, ate it within a few hours. Poduras feed ordinarily
on vegetable matter, such as dead leaves and growing crypto-
gamic vegetation. These little creatures can be easily preserved
in a mixture of alcohol and glycerine, or pure alcohol, though
without the glycerine the colors fade.

We have entered more fully in this chapter into the details of
structure than heretofore, too much so, perhaps, for the patience
of our readers. But the study of the Poduras possesses the
liveliest interest, since these lowest of all the six-footed insects
may have been among the earliest land animals, and hence to
them we may look with more or less success for the primitive,
ancestral forms of insect life.

# CHAPTER XIII.

## HINTS ON THE ANCESTRY OF INSECTS.

THOUGH our course through the different groups of insects may have seemed rambling and desultory enough, and pursued with slight reference to a natural classification of the insects of which we have spoken, yet beginning with the Hive bee, the highest intelligence in the vast world of insects, we have gradually, though with many a sudden step, descended to perhaps the most lowly organized forms among all the insects, the parasitic mites. While the Demodex is probably the humblest in its organization of any of the insects we have treated of, there is still another mite, which some eminent naturalists continue to regard as a worm, which is yet lower in the scale. This is the Pentastoma (Fig. 177, P. tænioides), which lives in the manner of the tape worm a parasitic life in the higher animals, though instead of inhabiting the alimentary canal, the worm-like mite takes up its abode in the nostrils and frontal

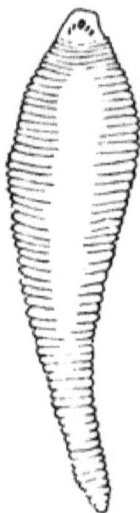

177. Pentastoma.

178. Centipede.

sinus of dogs and sheep, and sometimes of the horse. At first, however, it is found in the liver or lungs of various animals, sometimes in man. It is then in the earliest or larval state, and assumes its true mite form, being oval in shape, with minute horny jaws adapted for boring, and with two pairs of legs armed

(148)

with sharp retractile claws.    Such an animal as this is little
higher than some worms, and indeed is lower than many of them.

We should also not pass over in silence the Centipedes (Fig.
178, Scolopocryptops sexspinosa) and Galley worms, or Thousand legs and their allies (Myriopods), which by their long
slender bodies, and great number of segments and feet, vaguely
recall the worms.    But they, with the mites, are true insects, as
they are born with only three pairs of feet, as are the mites and
ticks, and breathe by tracheæ; and thus a common plan of
structure underlies the entire class of insects.

A very strange Myriopod has been discovered by Sir John

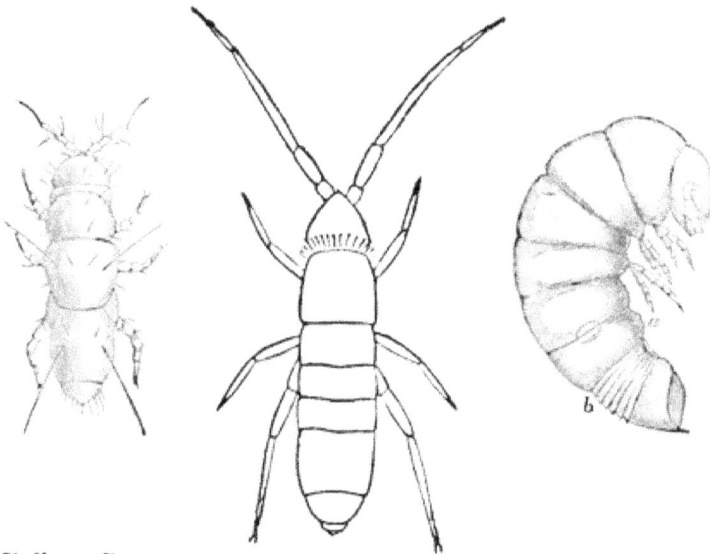

179. Young Pauropus.        180. Spring-tail.            181. Young Julus.

Lubbock in Europe, and we have been fortunate enough to find
a species in this country.    It is the Panropus.    It consists, when
fully grown, of nine segments, exclusive of the head, bearing
nine pairs of feet.    The young of Pauropus (Fig. 179) is born
with three pairs of feet, and in its general appearance reminds
us of a spring-tail (Fig. 180) as may be seen by a glance at the
cut.    This six-legged form of Panropus may also be compared
with the young galley worm (Fig. 181).

Passing to the group of spiders and mites, we find that the
young mites when first hatched have but three pairs of feet,
while their parents have four, like the spiders.    Figure 182

represents the larva (Leptus) of the red garden mites; while a figure of the "water bear," or Tardigrade (Fig. 183), is introduced to compare with it, as it bears a resemblance to the young of the mites, though their young are born with their full complement of legs, an exception to their nearest allies, the true mites. Now if we compare these early stages of mites and myriopods with those of the true six-footed insects, as in the larval Meloë, Cicada, Thrips and Dragon fly, we shall see quite plainly that they all share a common form. What does this mean? To the systematist who concerns himself with the classification of the myriads of different insects now living, it is a relief to find that all can be reduced to the

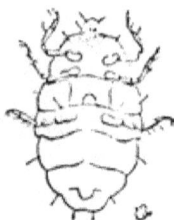

182. Leptus.

comparatively simple forms sketched above. It is to him a proof of the unity of organization pervading the world of insects. He sees how nature, seizing upon this archetypal form has, by simple modifications of parts here and there, by the addition of wings and other organs wanting in these simple creatures, rung numberless changes in this elemental form. And starting from the simplest kinds, such as the Poduras, Spiders, Grasshoppers and May flies, allied creatures which we now know were the first to appear in the earlier geologic ages, we rise to the highest, the bees with their complex forms, their diversified economy and wonderful instincts. In ascending this scale of being, while there is a progress upwards, the beetles, for instance, being higher than the bugs and

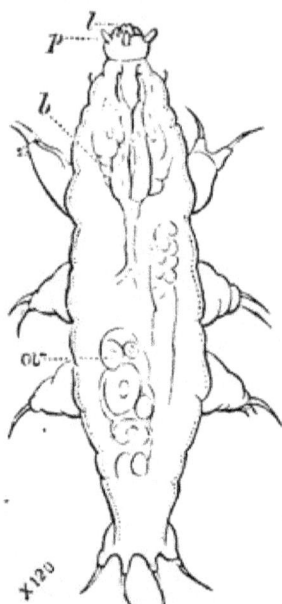

183. Tardigrade.

grasshoppers; and the butterflies and moths, on the whole, being more highly organized than the flies; and while we see the hymenopterous saw-flies, with their larvæ mimicking so closely the caterpillars of the butterflies, in the progress from the saw-flies up to the bees we behold a gradual loss of the lower saw-fly characters in the Cynips and Chalcid flies, and see in

the sand-wasps and true wasps a constant and accelerating likeness to the bee form. Yet this continuity of improving organi-
zations is often broken, and we often see insects which recall
the earlier and more elementary forms.

Again, going back of the larval period, and studying the insect in the egg, we find that nearly all the insects yet observed
agree most strikingly in their mode of growth, so that, for
instance, the earlier stages of the germ of a bee, fly or beetle,
bear a remarkable resemblance to each other, and suggest again,
more forcibly than when we examine the larval condition, that
a common design or pattern at first pervades all. In the light
of the studies of Von Baer, of Lamarck and Darwin, should we
be content to stop here, or does this ideal archetype become
endowed with life and
have a definite exis-
tence, becoming the
ancestral form of all
insects, the prototype
which gave birth to
the hundreds of thou-
sands of insect forms
which are now spread
over our globe, just
as we see daily hap-
pens where a single
aphis may become the

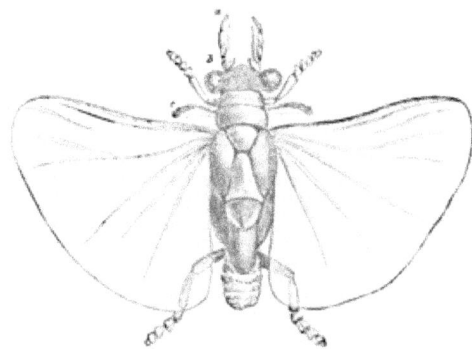

1-t. Male Stylops.

progenitor of a million offspring clustering on the same tree?
Is there not something more than analogy in the two things, and
is not the same life-giving force that evolves a million young
Aphides from the germ stock of a single Aphis in a single season, the same in kind with the production of the living races
of insects from a primeval ancestor? When we see the Aphis
giving origin in one season to successive generations, the individuals of which may be counted by the million, it is no less
mysterious than that other succession of forms of insect life
which has peopled the globe during the successive chapters of
its history. While we see in one case the origin of individual
forms, and cannot explain what it is that starts the life in the
germ and so unerringly guides the course of the growing embryo, it is illogical to deny that the same life-giving force is
concerned in the production of specific and generic forms.

Who can explain the origin of the sexes? What is the cause
that determines that one individual in a brood of Stylops, for
example (Fig. 184, male; Fig. 185, grub-like female in the body
of its host), shall be but a grub, living as a parasite in the
body of its host, while its fellow shall be winged and as free in
its actions as the most highly organized insect? It is no less
mysterious, because it daily occurs before our eyes. So perhaps
none the less mysterious, and no more discordant with known
natural laws may the law that governs the origin of species
seem to those who come after us. Certainly the present
attempts to discover that law, however fatuitous they may
seem to many, are neither illogical, nor, judging by the impetus

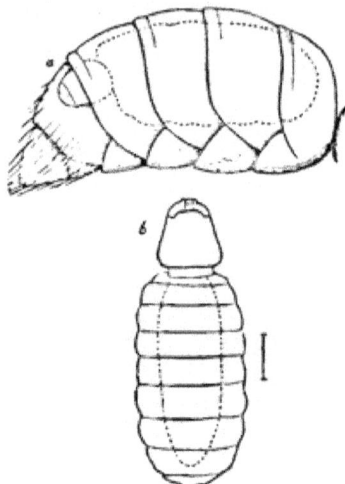

185. Female Stylops.

already given to biology, or the
science of life, labor altogether
spent in vain. The theory of
evolution is a powerful tool,
when judiciously used, that must
eventually wrest many a secret
from the grasp of nature.

But whether true or unproved,
the theory of evolution in some
shape has actually been adopted
by the large proportion of naturalists, who find it indispensable
in their researches, and it will be
used until found inadequate to
explain facts. Notwithstanding
the present distrust, and even
fear, with which it is received

by many, we doubt not but that in comparatively few years all
will acknowledge that the theory of evolution will be to biology
what the nebular hypothesis is to geology, or the atomic theory
is to chemistry. While the evolution theory is as yet imperfect,
and many objections, some seemingly insuperable, can be raised
against it, it should be borne in mind that the nebular hypothesis is still comparatively crude and unsatisfactory, though
indispensable as a working theory to the geologist; and in
chemistry, though the atomic theory may not be satisfactorily
demonstrated to some minds until an atom is actually brought
to sight, it is yet invaluable in research.

Many short sighted persons complain that such a theory sets

in the back-ground the idea of a personal Creator; but minds no less devout, and perhaps a trifle more thoughtful, see the hand of a Creator not less in the evolution of plants and animals from preexistent forms, through natural laws, than in the evolution of a summer's shower, through the laws discovered by the meteorologist, who looks back through myriads of ages to the causes that led to the distribution of mountain chains, ocean currents and trade winds, which combine to produce the necessary conditions resulting in that shower.

Indeed, to the student of nature, the evolution theory in biology, with the nebular hypothesis, and the grand law in physics of the correlation of forces, all interdependent, and revealing to us the mode in which the Creator of the Universe works in the world of matter, together form an immeasurably grander conception of the order of creation and its Ordainer, than was possible for us to form before these laws were discovered and put to practical use. We may be allowed, then, in a reverent spirit of inquiry, to attempt to trace the ancestry of the insects, and without arriving, perhaps, at any certain result, for it is largely a matter of speculation, point out certain facts, the thoughtful consideration of which may throw light on this difficult and embarrassing question.

Without much doubt the Poduras are the lowest of the six-footed insects. They are more embryonic in their appearance than others, as seen in the large size of the head compared with the rest of the body, the large, clumsy legs, and the equality in the size of the several segments composing the body. In other characters, such as the want of compound eyes, the absence of wings, the absence of a complete ovipositor, and the occasional want of tracheæ, they stand at the base of the insect series. That they are true insects, however, we endeavored to show in the previous chapter, and that they are neuropterous, we think is most probable, since not only in the structure of the insect after birth do they agree with the larvæ of certain neuropters, but, as we have shown in another place * in comparing the development of Isotoma, a Poduran, with that of a species of Caddis fly, the correspondence throughout the different embryological stages, nearly up to the time of hatching, is very striking. And it is a

---

* Memoirs of the Peabody Academy of Science, II. Embryological Studies on Diplax, Perithemis, and the Thysanurus genus Isotoma. Salem, 1871.

remarkable fact, as we have previously noticed, that when it begins to differ from the Caddis fly embryo, it begins to assume the Poduran characters, and its development consequently in some degree retrogrades, just as in the lice previous to hatching, as we have shown in a previous chapter, so that I think we are warranted at present in regarding the Thysanura, and especially the family of Podurids as degraded neuropters. Consequently the Poduras did not have an independent

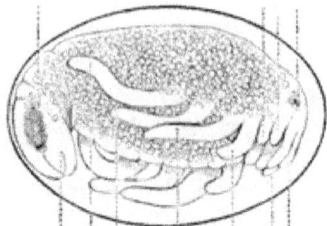

186. Embryo of Diplax.

origin and do not, perhaps, represent a distinct branch of the genealogical tree of articulates. While the Poduras may be said to form a specialized type, the Bristle-tails (Lepisma, Machilis, Nicoletia and Campodea) are, as we have seen, much more highly organized, and form a generalized or comprehensive type. They resemble in their general form the larva of Ephemerids, and perhaps more closely the immature Perla, and also the wingless cockroaches.

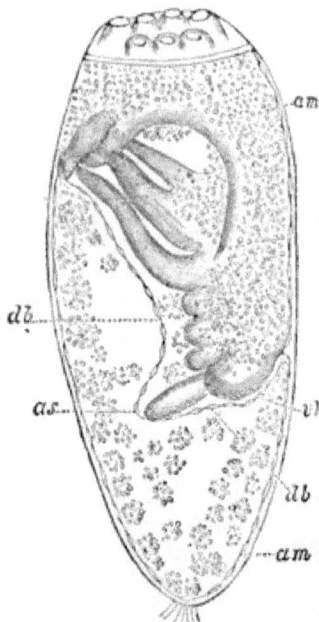

Now such forms as these Thysanura, together with the mites and the singular Pauropus, we cannot avoid suspecting to have been among the earliest to appear upon the earth, and putting together the facts, first, of their low organization; secondly, of their comprehensive structure, resembling the larvæ of other insects; and thirdly, of their probable great antiquity, we naturally look to them as being related in form to what we may conceive to have been the ancestor

187. Embryo of Louse.

of the class of insects. Not that the animals mentioned above were the actual ancestors, but that certain insects bearing a greater resemblance to them than any others with which we are acquainted, and belonging possibly to families and orders now extinct, were the prototypes and progenitors of the insects now known.

Though the study of the embryology of insects is as yet in its infancy, still with the facts now in our possession we can state with tolerable certainty that at first the embryos of all insects are remarkably alike, and the process of development is much the same in all, as seen in the figure of Diplax (Fig. 186), the louse (Fig. 187), the spider (Fig. 188) and the Podura (Fig. 189), and we could give others bearing the same likeness. We notice that at a certain period in the life of the embryo all agree in having the head large, and bearing from two to four pairs of mouth organs, resembling the legs; the thorax is merged in with the abdomen, and the general form of the embryo is

188. Embryo of Spider.

ovate. Now this general embryonic form characterizes the larva of the mites, of the myriopods and of the true insects. To such a generalized embryonic form to which the insects may be

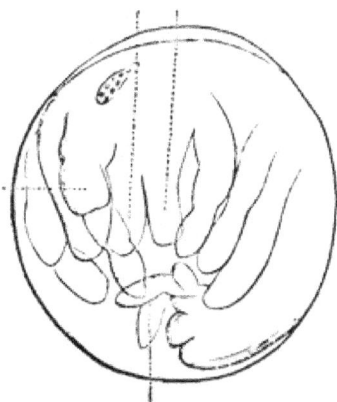

189. Embryo of Podura.

referred as the descendants, we would give the name of *Leptus*, as among Crustacea the ancestral form is referred to Nauplius, a larval form of the lower Crustacea, and through which the greater part of the Crabs, Shrimps, Barnacles, water fleas, etc., pass to attain their definite adult condition. A little water flea was described as a separate genus, Nauplius, before it was known to be the larva of a higher water flea, and so also Leptus was thought to be a mature mite. Accordingly, we follow the usage of certain naturalists in dealing with the Crustacea, and propose for this common primitive larval condition of insects the term Leptus.

The first to discuss this subject of the ancestry of insects was Fritz Müller, who in his " Für Darwin,"* published in 1863, says, at the end of his work, " Having reached the Nauplius, the extreme outpost of the class, retiring farthest into the gray mist of primitive time, we naturally look round us to see whether ways may not be descried thence towards other bordering regions. * * * But I can see nothing certain. Even towards the nearer provinces of the Myriopoda and Arachnida I can find no bridge. For the Insecta alone, the development of the Malacostraca [Crabs, Lobsters, Shrimps, etc.] may perhaps present a point of union. Like many Zoëæ, the Insecta possess three pairs of limbs serving for the reception of nourishment, and three pairs serving for locomotion; like the Zoëæ they have an abdomen without appendages; as in all Zoëæ the mandibles in Insecta are destitute of palpi. Certainly but little in common, compared with the much which distinguishes these two animal forms. Nevertheless, the supposition that the Insecta had for their common ancestor a Zoëa which raised itself into a life on land, may be recommended for further examination " (p. 140).

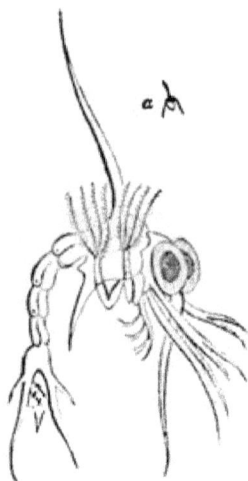

190. Zoëa.

Afterwards Hæckel in his "Generelle Morphologie" (1866) and "History of Creation," published in 1868, reiterates the notion that the insects are derived from the larva (Zoëa, Fig. 190) of the crabs, though he is doubtful whether they did not originate directly from the worms.†

It may be said in opposition to the view that the insects came

---

* Translated in 1869 by Mr. Dallas under the title "Facts for Darwin."

†"Whether that common stem-form of all the Tracheata [Insects, Myriopods and Spiders] which I have called Protracheata in my 'General Morphology' has developed directly from the true Annelides (Celelminthes), or, the next thing to this (zunachst), out of Zoea-form Crustacea (Zoepoda), will be hereafter established only through a sufficient knowledge and comparison of the structure and mode of growth of the Tracheata, Crustacea and Annelides. In either case is the root of the Tracheata, as also of the Crustacea, to be sought in the group of the true jointed worms (Annelides, Gephyrea and Rotatoria." He considers the first insect to have appeared after the Silurian period, viz., in the Devonian.

originally from the same early crustacean resembling the larva of a crab or shrimp, that the differences between the two types are too great, or, in other words, the homologies of the two classes too remote,[*] and the two types are each too specialized to lead us to suppose that one was derived from the other. Moreover, we find through the researches of Messrs. Hartt and Scudder that there were highly developed insects, such as May flies, grasshoppers, etc., in the Devonian rocks of New Brunswick, leading us to expect the discovery of low insects even in the Upper Silurian rocks. At any rate this discovery pushes back the origin of insects beyond a time when there were true Zoëæ, as the shrimps and their allies are not actually known to exist so far back as the Silurian, not having as yet been found below the coal measures.

The view that the insects were derived from a Zoëa was also sustained by Friedrich Brauer, the distinguished entomologist of Vienna, in a paper[†] read in March, 1869. Following the suggestion of Fritz Müller and Haeckel, he derives the ancestry of insects from the Zoëa of crabs and shrimps. However, he regards the Podurids as the more immediate ancestors of the true insects, selecting Campodea as the type of such an ancestral form, remarking that the "Campodea-stage has for the Insects and Myriopods the same value as the Zoëa for the Crustacea." He says nothing regarding the spiders and mites.

At the same time[‡] the writer, in criticising Haeckel's views of the derivation of insects from the Crustacea (ignorant of the fact that he had also suggested that the insects were possibly derived directly from the worms, and also independently of Brauer's opinions) declared his belief that though it seemed premature, after the discovery of highly organized winged insects

---

[*] The Zoëa is born with eight pairs of jointed appendages belonging to the head, and with no thoracic limbs, while in insects there are but four pairs of cephalic appendages and three pairs of legs. Correlated with this difference is the entirely different mode of grouping the body segments, the head and thorax being united into one region in the crab, but separate in the insects, the body being as a rule divided into a head, thorax and abdomen, while these regions are much less distinctly marked in the crabs, and liable in the different orders to great variations. The great differences between the Crustacea and insects are noticeable at an early period in the egg.

[†] Considerations on the Transmutation of Insects in the Sense of the Theory of Descent. Read before the Imperial Zoological-botanical Society in Vienna, April 3, 1869.

[‡] American Naturalist, vol. 3, p. 45. March, 1869.

14

in rocks so ancient as the Devonian, and with the late discovery
of a land plant in the Lower Silurian rocks of Sweden,* to even
guess as to the ancestry of insects, yet he would suggest that,
instead of being derived from some Zoëa, "the ancestors of the
insects (including the six-footed insects, spiders and myriopods)
must have been worm-like and aquatic, and when the type
became terrestrial we would imagine a form somewhat like the
young Pauropus, which combines in a remarkable degree the
characters of the myriopods and the degraded wingless insects,
such as the Smynthurus, Podura, etc. Some such forms may
have been introduced late in the Silurian period, for the interesting discoveries of fossil insects in the Devonian of New
Brunswick, by Messrs. Hartt and Scudder, and those discovered
by Messrs. Meek and Worthen in the lower part of the Coal
Measures at Morris, Illinois, and described by Mr. Scudder,
reveal carboniferous myriopods (two species of Euphorberia)
more highly organized than Pauropus, and a carboniferous scor-
pion (Buthus?) closely resembling a species now living in California, together with another scorpion-like animal, Mazonia

---

* See Prof. Torell's discovery of Eophyton Linnæanum, a supposed land plant
allied to the rushes and grasses of our day, in certain Swedish rocks of Lower
Cambrian age. The writer has, through the kindness of Prof. Torell, seen specimens of these plants in the Museum of the Geological Survey at Stockholm. Mr.
Murray, of the Canadian Geological Survey, was the first to discover in America
(Labrador, Straits of Belle Isle) this same genus of plants. They are described
and figured by Mr. Billings, who speaks of them as "slender, cylindrical, straight,
reed-like plants," in the "Canadian Naturalist" for August, 1872.

Should the terrestrial nature of these plants be established on farther evidence,
then we are warranted in supposing that there were isolated patches of land in
the Cambrian or Primordial period, and if there was land there must have been
bodies of fresh water, hence there may have been both terrestrial and aquatic
insects, possibly of forms like the Podurids, May flies, Perlæ, mites and Pauropus
of the present day. There was at any rate land in the Upper Silurian period, as
Dr. J. W. Dawson describes land plants (Psilophyton) from the Lower Helderberg
Rocks of Gaspe, New Brunswick, corresponding in age with the Ludlow rocks of
England.

We might also state in this connection that Dr. Dawson, the eminent fossil botanist of Montreal, concludes from the immense masses of carbon in the form of
graphite in the Laurentian rocks of Canada, that "the Laurentian period was
probably an age of most prolific vegetable growth. * * * Whether the vegetation of the Laurentian was wholly aquatic or in part terrestrial we have no means
of knowing." In 1855, Dr. T. Sterry Hunt asserted "that the presence of iron ores,
not less than that of graphite, points to the existence of organic life even during
the Laurentian or so-called Azoic period." In 1861 he went farther and stated his
belief in "the existence of an abundant vegetation during the Laurentian period."
The Eophyton in Labrador occurs above the Trilobite (Paradoxides) beds, while
in Sweden they occur below.

Woodiana, while the Devonian insects described from St. John by Mr. Scudder, are nearly as highly organized as our grasshoppers and May flies. Dr. Dawson has also discovered a well developed milleped (Xylobius) in the Lower Coal Measures of Nova Scotia; so that we must go back to the Silurian period in our search for the earliest ancestor, or (if not of Darwinian proclivities) prototype, of insects."

Afterwards* the writer, carrying out the idea suggested above, "referred the ancestry of the Myriopods, Arachnids, and Hexapodous insects to a Leptus-like terrestrial animal, bearing a vague resemblance to the Nauplius form among Crustacea, inasmuch as the body is not differentiated into a head, thorax and abdomen [though the head may be free from the rest of the body] and there are three pairs of temporary locomotive appendages. Like Nauplius, which was first supposed to be an adult Entomostracan, the larval form of Trombidium had been described as a genus of mites under the name of Leptus (also Ocypete and Astoma) and was supposed to be adult."

In the same year Sir John Lubbock† agrees with Brauer that the groups represented by Podura and Campodea may have been the ancestors of the insects, remarking that "the genus Campodea must be regarded as a form of remarkable interest, since it is the living representative of a primæval type from which not only the Collembola (Podura, etc.) and Thysanura, but the other great orders of insects, have all derived their origin."

The comparison of the Leptus with the Nauplius, or pre-Zoëal stage of Crustacea, is much more natural. But here we are met with apparently insuperable difficulties. While the Nauplius (Fig. 194) has but three pairs of appendages, which become the two pairs of antennæ and succeeding pair of limbs of the adult, in the Leptus as the least number we have five pairs, two of which belong to the head (the maxillæ and mandibles) and three to the thorax; besides these is a true head, distinct from the hinder region of the body. It is evident that the Leptus fundamentally differs from the Nauplius and begins life on a higher plane. We reject, therefore, the Crustacean origin of the insects. Our only refuge is in the worms, and how to account

<hr/>

* In a communication made to the Boston Society of Natural History, Oct. 17, 1870 (see also "American Naturalist" for Feb. and Sept., 1871).

† On the Origin of Insects, a paper read before the Linnæan Society of London Nov. 2, 1871, and reported in abstract in "Nature," Nov. 9, 1871.

for the transmutation of any worm with which we are at present
acquainted into a form like the Leptus, with its mandibulated
mouth and jointed legs, seems at first well nigh impossible.
We have the faintest possible indication in the structure of
some mites, and of the Tardigrades and Pentastoma, where
there is a striking recurrence, as we may term it, to a worm-like
form, readily noticed by every observer, whatever his opinion
may be on the developmental theory. In the Demodex we see
a tendency of the mite to assume under peculiar circumstances
an elongated, worm-like form. The mouth-parts are aborted
(though from what we know of the embryology of other mites,

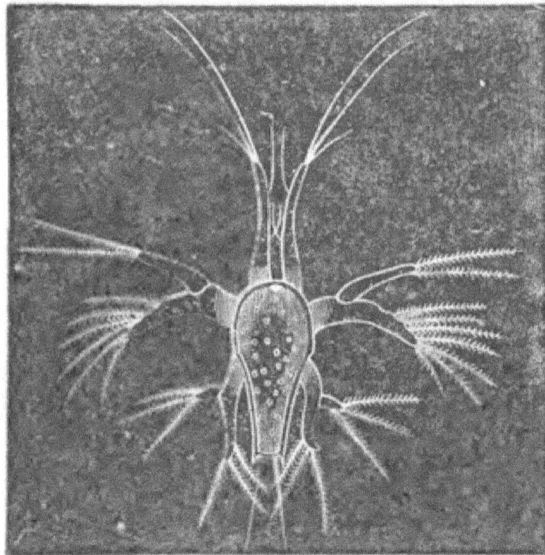

191. Nauplius.

they probably
are indicated
early in embry-
onic life), while
the eight legs are
not jointed, and
form simple tu-
bercles. In the
Tardigrades, a
long step lower,
we have un-
jointed fleshy
legs armed with
from two to four
claws, but the
mouth-parts are
essentially mite
in character. A
decided worm
feature is the fact that they are hermaphrodites, each individual
having ovaries and spermaries, as is the case with many worms.

When we come to the singular creatures of which Pentastoma
and Linguatula are the type, we have the most striking approximation to the worms in external form, but these are induced
evidently by their parasitic mode of life. They lose the rudimentary jointed limbs which some (Linguatula especially) have
well marked in the embryo, and from being oval, rudely mite-
like in form, they elongate, and only the claws or simple curved
hooks, like those of young tape worms, remain to indicate the
original presence of true jointed legs.

In seeking for the ancestry of our hypothetical Leptus among the worms, we are at best groping in the dark. We know of no ancestral form among the true Annelides, nor is it probable that it was derived from the intestinal worms. The only worm below the true Annelides that suggests any remote analogy to the insects is the singular and rare Peripatus, which lives on land in warm climates. Its body, not divided into rings, is provided with about thirty pairs of fleshy tubercles, each ending in two strong claws, and the head is adorned with a pair of fleshy tubercles. It is remotely possible that some Silurian land worm, if any such existed, allied to our living Peripatus, may have been the ancestor of a series of types now lost which resulted in an animal resembling the Leptus.

We may, however, as bearing upon this difficult question, cite some remarkable discoveries of Professor Ganin, a Russian naturalist, on the early stages of certain ichneumon parasites, which show some worm features in their embryonic development. In a species of Platygaster (Fig. 192, P. error of Fitch), which is a parasite on a two-winged gall fly, the earliest stage observed after the egg is laid is that in which the egg contains a single cell with a

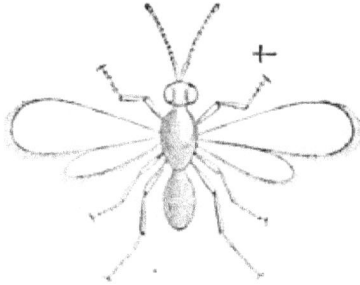

192. Platygaster error.

nucleus and nucleolus. Out of this cell (Fig. 193 A, a) arise two other cells. The central cell (a) gives origin to the embryo. The two outer ones multiply by subdivision and form the embryonal membrane, or "amnion," which is a provisional envelope and does not assist in building up the body of the germ. The central single cell, however, multiplies by the subdivision of its nucleus, thus building up the body of the germ. Figure 193 B, g, shows the yolk or germ just forming out of the nuclei (a) and b, the peripheral cells of the blastoderm skin, or "amnion." Figure 193 C shows the yolk transformed into the embryo (g), with the outer layer of blastodermic cells (b). The body of the germ is infolded, so that the embryo appears bent on itself. Figure 193 D shows the embryo much farther advanced, with the two pairs of lobes (md, rudimentary mandibles; d, rudimentary pad-like organs, seen in a more

advanced stage in $E$), and the bilobate tail ($st$).    Figure 194 ($m$,
mouth; $at$, rudimentary antennæ; $md$, mandibles; $d$, tongue-like
appendages; $st$, anal stylets; the subject of this figure is of a
different species from the insect previously figured, which, however, it closely resembles) shows the first larva stage after
leaving the egg.    This strange form, the author remarks, would
scarcely be thought an insect, were not its origin and farther
development known, but rather a parasitic Copepodous crustacean, whence he calls this the Cyclops-like stage.    In this con-

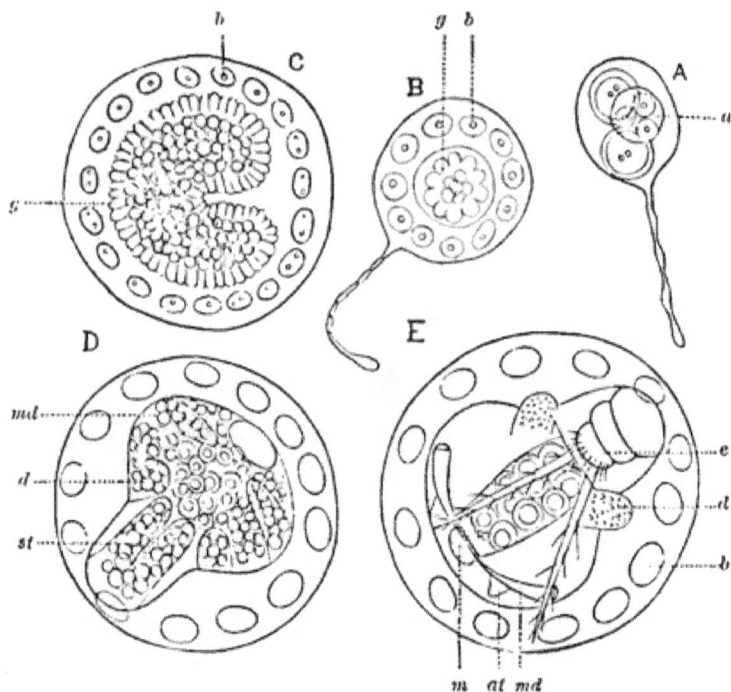

193. Development of Platygaster.

dition it clings to the inside of its host by means of its hook-like
jaws ($md$), moving about like a Cestodes embryo with its well
known six hooks.    The tail moves up and down, and is of but
little assistance in its efforts to change its place.    Singularly
enough, the nervous, vascular, and respiratory systems (tracheæ) are wanting, and the alimentary canal is a blind sac,
remaining in an indifferent, or unorganized state.    How long
it remains in this state could not be ascertained.

The second larval stage (Fig. 195; _œ_, œsophagus; _ng_, supra-œsophageal ganglion; _n_, nervous cord; _ga_, and _g_, genital organs; _ms_, band of muscles) is attained by means of a moult, as usual in the metamorphoses of insects. With the change of skin the larva entirely changes its form. So-called hypodermic cells are developed. The singular tail is dropped, the segments of the body disappear, and the body grows oval, while within begins

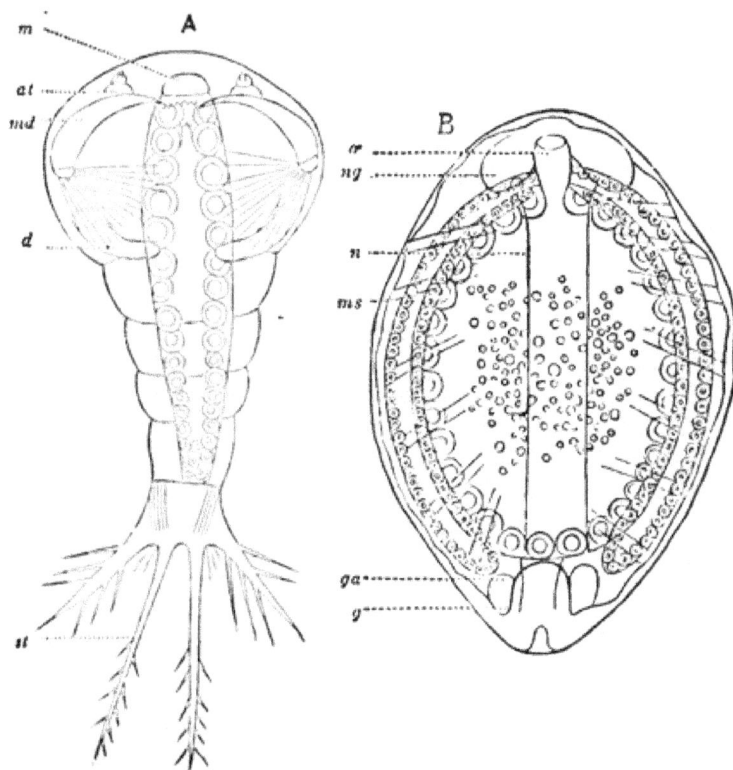

191. First Larva of Platygaster.    195. Second Larva of Platygaster.

a series of remarkable changes, like the ordinary development of the embryo of most other insects within the egg. The cells of the hypodermis multiply greatly, and lie one above the other in numerous layers. They give rise to a special primitive organ closely resembling the "primitive band" of all insect embryos. The alimentary canal is made anew, and the nervous and vascu-

lar systems now appear, but the tracheæ are not yet formed.
It remains in this state for a much longer period than in the
previous stage.

The third larval form only a few live to reach. This is of the
usual long, oval form of the larvæ of the ichneumons, and the
body has thirteen segments exclusive of the head. The muscular
system has greatly developed and the larva is much more lively
in its motions than before. The new
organs that develop are the air tubes and
fat bodies. The "imaginal disks" or rudimentary portions destined to develop and
form the skin of the adult, or imago, arise
in the pupa state, which resembles that of
other ichneumons. These disks are only
engaged, in Platygaster, in building up
the rudimentary appendages, while in the
flies (Muscidæ and Corethra) they build
up the whole body, according to the
remarkable discovery of Weismann.

Not less interesting is the history of the
development of a species of Polynema,
another egg-parasite, which lays its eggs
(one, seldom two) in the eggs of a small
dragon fly, Agrion virgo, which oviposits
in the parenchyma of the leaves of water-
lilies. The eggs develop as in Platygaster.
The earliest stage of the embryo is very
remarkable. It leaves the egg when very
small and immovable, and with scarcely a
trace of organization, being a mere flask-shaped sac of cells.*
It remains in this state five or six days.

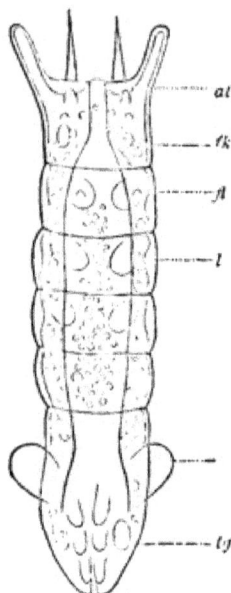

196. Third Larva of
Polynema.

In the second stage, or Histriobdella-like form, the larva is,
in its general appearance, like the low worm to which Ganin
compares it. It may be described as bearing a general resemblance to the third and fully developed larval form (Fig. 196, *tg*,

---

* This reminds us (though Ganin does not mention it) of the development of the
embryo of Julus, the Thousand legs, which, according to Newport, hatches the
25th day after the egg is laid. At this period the embryo is partially organized,
having faint traces of segments, and is still enveloped in its embryonal membranes
and retains its connection with the shell. In this condition it remains for seventeen days, when it throws off its embryonal membrane, and becomes detached
from the shell.

three pairs of abdominal tubercles destined to form the sting; *l*, rudiments of the legs; *fk*, portion of the fatty body; *at*, rudiments of the antennæ; *fl*, imaginal disks, or rudiments of the wings). No tracheæ are developed in the larva, nor do any exist in the imago. (Ganin thinks, that as these insects are somewhat aquatic, the adult insects flying over the surface of the water, the wings may act as respiratory organs, like gills.) It lives six to seven days before pupating, and remains from ten to twelve days in the pupa state.

The origin of the sting is clearly ascertained. Ganin shows that it consists of three pairs of tubercles, situated respectively on the seventh, eighth, and ninth segments of the abdomen (Fig. 196, *ly*). The labium is not developed from a pair of tubercles, as is usual, but at once appears as an unpaired, or single organ. The pupa state lasts for five or six days, and

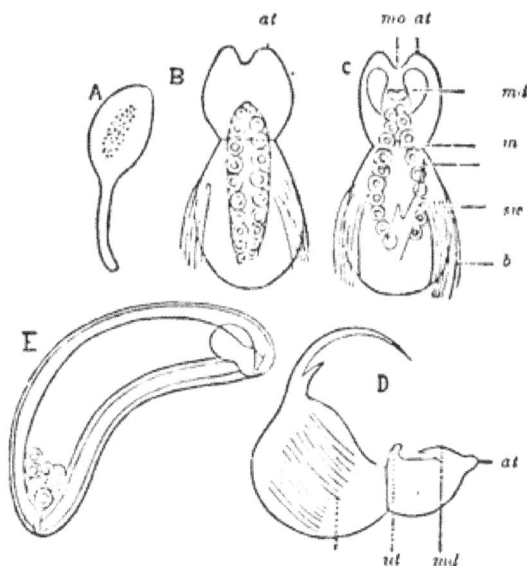

197. Development of Egg-parasites.

when the imago appears it eats its way through a small round opening in the end of the skin of its host, the Agrion larva.

The development of Ophioneurus, another egg-parasite, agrees with that of Platygaster and Polynema. This egg-parasite passes its early life in the eggs of Pieris brassicæ, and two or three live to reach the imago state, though about six eggs are deposited by the female. The eggs are oval, and not stalked. The larva is at first of the form indicated by figure 197 *E*, and when fully grown becomes of a broad oval form, the body not being divided into segments. It differs from the genera already mentioned, in remaining within its egg membrane, and not assuming their strange *forms.* From the non-segmented, sac-like larva, it passes directly into the pupa state.

The last egg-parasite noticed by Ganin, is Teleas, whose development resembles that of Platygaster. It is a parasite in the eggs of Gerris, the Water Boatman. Figure 197 *A* represents the egg; *B, C,* and *D,* the first stage of the larva, the abdomen (or posterior division of the body) being furnished with a series of bristles on each side. (*B* represents the ventral, *C* the dorsal, and *D* the profile view; *at,* antennae; *md,* hook-like mandibles; *mo,* mouth; *b,* bristles; *m,* intestine; *sw,* the tail; *ul,* under lip or labium.) In the second larval stage, which is oval in form, and not segmented, the primitive band is formed.

In concluding the account of his remarkable discoveries, Ganin draws attention to the great differences in the formation of the eggs and the germs of these parasites from what occurs in other insects. The egg has no nutritive cells; the formation of the primitive band, usually the first indication of the germ, is retarded till the second larval stage is attained; and the embryonal membrane is not homologous with the so-called "amnion" of other insects, but may possibly be compared with the skin developed on the upper side of the low, worm-like acarian, Pentastomum, and the "larval skin" of the embryos of many low Crustacea. He says, also, that we cannot, perhaps, find the homologues of the provisional organs of the larvæ, such as the singularly shaped antennæ, the claw-like mandibles, the tongue- or ear-like appendages, in other Arthropoda (insects and Crustacea); but that they may be found in the parasitic Lernæan crustaceans, and in the leeches, such as Histriobella. He is also struck by the similarity in the development of these egg-parasites to that of a kind of leech (Nephelis), the embryo of which is provided with ciliæ, recalling the larva of Teleas (Fig. 197 *B, C*), while in the true leeches (Hirudo) the primitive band is not developed until after they have passed through a provisional larval stage.

This complicated metamorphosis of the egg-parasites, Ganin also compares to the so-called "hyper-metamorphosis" of certain insects (Meloe, Sitaris, and the Stylopidæ) made known by Siebold, Newport and Fabre, and he considers it to be of the same nature.

He also, in closing, compares such early larval forms as those given in figures 193 *E* and 194, to the free swimming Copepoda. Finally, he says a few words on the theory of evolution, and remarks "there is no doubt that, if a solution of the questions arising concerning the genealogical relations of different animals among themselves is possible, comparative embryology will afford the first and truest principles." He modestly suggests that the facts presented in his paper will widen our views on the genetic relations of the insects to other animals, and refers to the opinion first expressed by Fritz Müller (Für Darwin, p. 91), and endorsed by Haeckel in his "Generelle Morphologie," that we must seek for the ancestors of insects and Arachnida in the Zoëa form of Crustacea. He cautiously remarks, however, that "the embryos and larvæ observed by me in the egg-parasites open up a new and wide field for a whole series of such considerations; but I will suppress them, since I am firmly convinced that a theory, which I build up to-day, can easily be destroyed with some few facts which I learn to-morrow. Since comparative embryology as a science does not yet exist, so do I think that all genetic theories are too premature, and without a strong scientific foundation."

The writer is perhaps less cautious, but he cannot refrain from making some reflections suggested by the remarkable discoveries of Ganin. In the first place, these facts bear strongly on the theory of evolution by "acceleration and retardation." In the history of these early larval stages we see a remarkable acceleration in the growth of the embryo. A simple sac of unorganized cells, with a half-made intestine, so to speak, is hatched, and made to perform the duty of an ordinary, quite highly organized larva. Even the formation of the "primitive band," usually the first indication of the organization of the germ, is postponed to a comparatively late period in larval life. The different anatomical systems, *i.e.*, the heart with its vessels, the nervous system and the respiratory system (tracheæ), appear at longer or shorter intervals, while in one genus the tracheæ are not developed at all. Thus some portions of the animal are accelerated in their development more than others, while others are retarded, and in some species certain organs are not developed at all. Meanwhile all live in a fluid medium, with much the same habits, and surrounded with quite similar physical conditions.

The highest degree of acceleration is seen in the reproductive organs of the Cecidomyian larva of Miastor, which produces a summer brood of young, alive, and living free in the body of the child-parent; and in the pupa of Chironomus, which has been recently shown by Von Grimm, a fellow countryman of Ganin, to produce young in the spring, while the adult fly lays eggs in the autumn in the usual manner. This is in fact a true virgin reproduction, and directly comparable to the alternation of generations observed in the jelly fishes, in Salpa, and certain intestinal worms. We can now, in the light of the researches of Siebold, Leuckart, Ganin and others, trace more closely than ever the connection between simple growth and metamorphosis, and metamorphosis and parthenogenesis, and perceive that they are but the terms of a single series. By the acceleration in the development of a single set of organs (the reproductive), no more wonderful than the acceleration and retardation of the other systems of organs, so clearly pointed out in the embryos of Platygaster and its allies, we see how parthenogenesis under certain conditions may result. The barren Platygaster larva, the fertile Cecidomyia larva, the fertile Aphis larva, the fertile Chironomus pupa, the fertile hydroid polype, and the fertile adult queen bee are simply animals in different degrees of organization, and with reproductive systems differing not in quality, but in the greater or less rapidity of their development as compared with the rest of the body.

Another interesting point is, that while the larvæ vary so remarkably in form, the adult ichneumon flies are remarkably similar to one another. Do the differences in their larval history seem to point back to certain still more divergent ancestral forms?

These remarkable hyper-metamorphoses remind us of the metamorphosis of the embryo of Echinoderms into the Pluteus-and Bipinnaria-forms of the starfish, sea urchins and Holothurians;* of the Actinotrocha-form larva of the Sipunculoid worms;

---

* It is a suggestive fact that these deciduous forms give way through histolysis to true larval forms, just as in some flies (Musca vomitoria) the true larval form goes under, and the adult form is built up from the imaginal disks of the larva. In an analogous manner the deciduous, pluteus-condition of the young Echinoderm perishes and is absorbed by the growing body of the permanent adult stage. This deciduous stage of the ichneumon may accordingly be termed the prelarval stage. Now as we find insects with and without this prelarval stage, and in the radiates quite different degrees of metamorphoses, the inquiry arises how far these differences are correlated with, and consequently dependent upon, the physical sur-

of the Tornaria into Balanoglossus, the worm; of the Cercaria-form larva of Distoma; of the Pilidium-form larva of Nemertes; and the larval forms of the leeches;* as well as the mite Pentastomum, and certain other aberrant mites, such as Myobia.

While Fritz Müller and Dohrn have considered the insects as having descended from the Crustacea (some primitive zoëa-form), and Dohrn has adduced the supposed zoëa-form larva of these egg-parasites as a proof, we cannot but think, in a subject so purely speculative as the ancestry of animals, that the facts brought out by Ganin tend to confirm our theory, that the ancestry of all the insects (including the Arachnids and Myriopods) should be traced directly to the worms.    The development of the degraded, aberrant Arachnidan Pentastomum accords, in some important respects, with that of the intestinal worms. The Leptus-form larva of Julus, with its strange embryological development, in some respects so like that of some worms, points in that direction, as certainly as does the embryological development of the egg-parasite Ophioneurus.    The Nauplius form of the embryo or larva of nearly all Crustacea, also points back to the worms as their ancestors, the divergence having perhaps originated, as we have suggested, in the Rotatoria.

While the Crustacea may have resulted from a series of prototypes leading up from the Rotifers (Fig. 198), it is barely

roundings of these animals in the free swimming condition.  Merely to point out the differences in the mode of development of animals is an interesting matter, and one could do worse things, but the philosophical naturalist cannot rest here. He must seek how these differences were brought about.

* Leuckart, in his great work, " Die Menschlichen Parasiten," p. 700, after the analogy of Hirudo, which develops a primitive streak late in larval life, ventures to consider the first indications of the germ of Nemertes in its larval, Pilidium form as a primitive streak.  He also suggests that the development of the later larval forms of the Echinoderms is the same in kind.

Moreover, nearly twenty years ago (1854) Zaddach, a German naturalist, contended that the worms are closely allied in their mode of development to the insects and crustaceans.  He compares the mode of development of a leech (Clepsine) and certain bristle-bearing worms (Saennris, Lumbriculus and Uaxes), and we may now from Kowalewsky's researches (1871) add the common earth worm (Lumbricus), in which there is no such metamorphosis as in the sea Nereids, to that of insects; the mode of formation of the primitive band in the leeches and earth worms being much like that of insects.  This confirms the view of Leuckart and Ganin, who both seem to have overlooked Zaddach's remarks.  Moreover, the rings of the harder bodied worms, as Zaddach says, contain chitine, as in the insects.  Zaddach also enters into farther details, which in his opinion ally the worms nearer to the insects than many naturalists at his time were disposed to allow.  The singular Echinoderes has some remarkable Arthropod characters.

15

possible that one of these creatures may have given rise to a form resulting in two series of beings, one leading to the Leptus form, the other to the Nauplius. For the true Annelides (Chætopods) are too circumscribed and homogeneous a group to allow us to look to them for the ancestral forms of insects. But that the insects may have descended from some low worms is not improbable when we reflect that the Syllis and allied genera of Annelides bear appendages consisting of numerous joints; indeed, the strange Dujardinia rotifera, figured by Quatrefages, in its general form is remarkably like the larva of

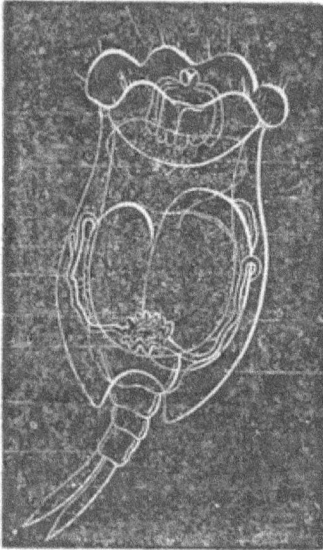

Chloëon. It has a quite distinct head, bearing five long, slender, jointed antennæ, and but eight or nine rings to the body, which ends in two long, many jointed appendages exactly like the tentacles. Quatrefages adds, that its movements are usually slow, but "when it wishes to move more rapidly, it moves its body alternately up and down with much vivacity, and shoots forwards by bounds, so to speak, a little after the manner of the larvæ of the mosquito" (Histoire Naturelle des Annclés, Tome 2, p. 69). The gills of aquatic insects only differ from those of worms in possessing trachcæ, though the gills of the Crustacea

198. A Rotifer.

may be directly compared with those of insects.

But when once inside the circle of the class of insects the ground is firmer, as our knowledge is surer. Granting now that the Leptus-like ancestor of the six-footed insects has become established, it is not so difficult to see how the Poduræ and finally a form like Campodea appeared. Aquatic forms resembling the larva of the Ephemeræ, Perlæ and, more remotely, the Forficulæ and white ants of to-day were probably evolved with comparative suddenness. Given the evolution of forms like the earwigs (Forficula), cockroaches and white ants (Termes), the latter of which abounded in the coal period, and it was not a great step forward to the evolution of the Dragonflies, the Psocus, the Chrysopa, the lice or parasitic Hemiptera, together with Thrips, thus forming the establishment of lines of development leading up to those Neuroptera with a complete metamorphosis, and finally to the grasshoppers and other forms of Orthoptera, together with the Hemiptera.

We have thus advanced from wingless to winged forms, *i. e.*,

 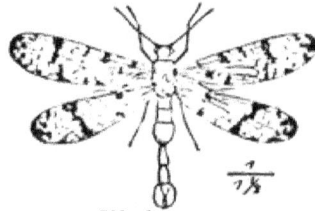

199. Chrysopa. 200. Panorpa.

from insects without a metamorphosis to those with a partial metamorphosis like the Perlas; to the May flies and Dragon flies, in which the adult is still more unlike the larva; to the Chrysopa (Fig. 199) and Forceps Tails (Panorpa. Fig. 200) and Caddis flies, in which, especially the latter, the metamorphosis is complete, the pupa being inactive and enclosed in a cocoon.

Having assumed the creation of our Leptus by evolutional laws, we must now account for the appearance of tracheæ and those organs so dependent on them, the wings, which, by their presence and consequent changes in the structure of the crust of the body, afford such distinctive characters to the flying insects, and raise them so far above the creeping spiders, and centipedes. Our Leptus at first undoubtedly breathed through the skin, as do most of the Poduras, since we have been unable to find tracheæ in them, nor even in the prolarva of a genus of minute ichneumon egg parasites, nor in the Linguatulæ and Tardigrades, and some mites, such as the Itch insect and the Demodex, and other Acari. In the Myriopod, Pauropus, Lubbock was unable to find any traces of tra-

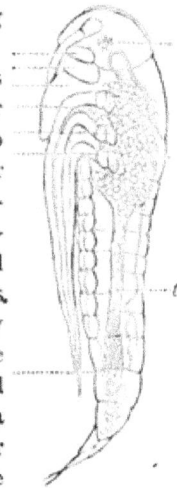

201. Embryo of Diplax.

cheæ. If we examine the embryo of an insect shortly before birth, as in the young Dragon fly (figure 201, the dotted line *t* crosses the rudimentary tracheæ), we find it to consist of

two simple tubes with few branches, while there are no stigmata, or breathing holes, to be seen in the sides of the body.
This fact sustains the view of Gegenbaur* that at first the tracheæ formed two simple tubes in the body-cavity, and that the
primary office of these tubes was for lightening the body, and
that their function as respiratory tubes was a secondary one.
The aquatic Protoleptus, as we may term the ancestor of Leptus, may have had such tubes as these, which acted like the
swimming bladder of fishes for lightening the body, as suggested
by Gegenbaur. It is known that the swimming bladder of fishes
becomes developed into the lungs of air-breathing vertebrates
and man himself. As our Leptus adopted a terrestrial life and
needed more air, a connection was probably formed by a minute
branch on each side of the body with some minute pore (for
such exist, whose uses are as yet unknown) through the skin,
which finally became specialized into a stigma, or breathing
pore ; and from the tracheal system being closed, we now have
the open tracheal system of land insects.

The next inquiry is as to the origin of the wings. Here the
question arises if wingless forms are exceptional among the
winged insects, and the loss of wings is obviously dependent
on the habits (as in the lice), and environment of the species
(as in beetles living on islands, which are apt to lose the hinder
pair of wings), why may not their acquisition in the first place
have been due to external agencies; and, as they are suddenly
discarded, why may they not have suddenly appeared in the first
place? In aquatic larvæ there are often external gill-like organs,
being simple sacs permeated by tracheæ (as in Agrion, Fig. 129,
or the May flies). These organs are virtually aquatic wings,
aiding the insect in progression as well as in aërating the blood,
as in the true wings. They are very variable in position, some
being developed at the extremity of the abdomen, as in Agrion,
or along the sides, as in the May flies, or filiform and arranged
in tufts on the under side of the body, as in Perla; and the naturalist is not surprised to find them absent or present in accordance with the varying habits of the animal. For example, in the
larvæ of the larger Dragon flies (Libellula, etc.) they are wanting, while in Agrion and its allies they are present.

---

*Vergleichende Anatomie, 2te Auflage, 1870, p. 437. I should, however, here add
that I am told by Mr. Putnam that some fishes which have no swim-bladder, are
surface-swimmers, and vice versa.

Now we conceive that wings formed in much the same way, and with no more disturbance, so to speak, to the insect's organization, appeared during a certain critical period in the metamorphosis of some early insect. As soon as this novel mode of locomotion became established we can easily see how surrounding circumstances would favor their farther development until the presence of wings became universal. If space permitted us to pursue this interesting subject farther, we could show how invariably correlated in form and structure are the wings of insects to the varied conditions by which they are surrounded, and which we are forced to believe stand in the relation of cause to effect. Again, why should the wings always appear on the thorax and on the upper instead of the under side? As this is the seat of the centre of gravity, it is evident that cosmical laws as well as the more immediate laws of biology determine the position and nature of the wings of an insect.

Correlated with the presence of wings is the wonderful differentiation of the crust, especially of the thorax, where each segment consists of a number of distinct pieces; while in the spiders and Myriopods the segments are as simple as in the abdominal segments of the winged insect. It is not difficult here to trace a series leading up from the Poduras, in which the segments are like those of spiders, to the wonderful complexity of the parts in the thoracic segments of the Lepidoptera and Hymenoptera.

In his remarks "On the Origin of Insects,"[*] Sir John Lubbock says, "I feel great difficulty in conceiving by what natural process an insect with a suctorial mouth like that of a gnat or butterfly could be developed from a powerfully mandibulate type like the Orthoptera, or even from the Neuroptera." Is it not more difficult to account for the origin of the mouth-parts at all? They are developed as tubercles or folds in the tegument, and are homologous with the legs. Figure 186 shows that the two sorts of limbs are at one time identical in form and relative position. The thought suggests itself that these long, soft, finger-like appendages may have been derived from the tentacles of the higher worms, but the grounds for this opinion are uncertain. At any rate, the earliest form of limb must have been that of a soft tubercle armed with one, or two, or many terminal

---

[*] Reported in " Nature " for Nov. 9, 1871.

claws, as seen in aquatic larvæ, such as Chironomus (Fig. 202), Ephydra (Fig. 203 *a*, *b*, *c*, pupa) and many others. As the Proto-leptus assumed a terrestrial life and needed to walk, the rudi-mentary feet would tend to elongate, and in consequence need the presence of chitine to harden the integument, until the habit of walking becoming fixed, the necessity of a jointed structure arose. After this the different needs of the offspring of such an

202. Foot of Chiro-nomus.

insect, with their different modes of taking food, vegetable or animal, would induce the diverse forms of simple, or raptorial, or leaping or digging limbs. A peculiar use of the anterior members, as seen in grasping the food and con-veying it to the mouth (perhaps originally a simple orifice with soft lips, as in Peripatus), would tend to cause such limbs to be grouped together, to concentrate around the mouth-opening, and to be directed constantly forwards. With use, as in the case of legs, these originally soft mouth-feet would gradually harden at the extremities, until serviceable in biting, when they would become jaws and palpi. Given a mouth and limbs surrounding it, and we at once have a rude head set off from the rest of the body. And in fact such is the history of the development of these parts in the embryo. At first the head is indicated by the buds forming the rudiments of limbs; the segments to which they are attached do not form a true head until after the mouth-parts have attained their jaw-like characters, and it is not un-til the insect is about to be hatched, that the head is definitely walled in.

We have arrived, then, at our Lep-tus, with a head bearing two pairs of jaws. The spiders and mites do not advance beyond this stage. But

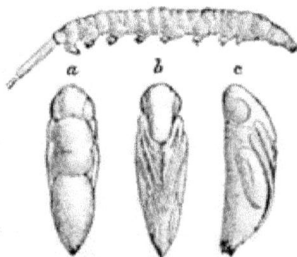

203. Ephydra.

in the true insects and Myriopods, we have the addition of special sense organs, the antennæ, and another pair of appen-dages, the labial palpi. It is evident that in the ancestor of these two groups the first pair of appendages became early adapted for purely sensory purposes, and were naturally pro-jected far in advance of the mouth, forming the antennæ.

Before considering the changes from the mandibulate form

of insects to those with mouth parts adapted for piercing and sucking, we must endeavor to learn how far it was possible for the caterpillar or maggot to become evolved from the Leptus-like larvæ of the Neuroptera, Orthoptera, Hemiptera and most Coleoptera. I may quote from a previous article* a few words in relation to two kinds of larvæ most prevalent among insects. "There are two forms of insectean larvæ which are pretty constant. One we call *leptiform*, from its general resemblance to the larvæ of the mites (Leptus). The larvæ of all the Neuroptera, except those of the Phryganeidæ and Panorpidæ (which are cylindrical and resemble caterpillars), are more or less leptiform, *i. e.*, have a flattened or oval body, with large thoracic legs. Such are the larvæ of the Orthoptera and Hemiptera, and the Coleoptera (except the Curculionidæ; possibly the Cerambycidæ and Buprestidæ, which approach the maggot-like form of the larvæ of weevils). On the other hand, taking the caterpillar or bee larva, with their cylindrical, fleshy bodies, in most respects typical of larval forms of the Hymenoptera, Lepidoptera and Diptera, as the type of the *eruciform* larva, etc. * * * The larvæ of the earliest insects were probably leptiform, and the eruciform condition is consequently an acquired one, as suggested by Fritz Müller."† It seems that these two sorts of larvæ had also been distinguished by Dr. Brauer in the article already referred to, with which, however, the writer was unacquainted at the time of writing the above quoted article. The similar views presented may seem to indicate that they are founded in nature. Dr. Brauer, after remarking that the Podurids seemed to fulfil Hæckel's idea of what were the most primitive insects, and noticing how closely they resemble the larvæ of Myriopods, says, "specially interesting are those forms among the Poduridæ which are described as Campodea and Japyx, since the larvæ of a great number of insects may be traced back to them"; but he adds, and with this view we are unable to agree, "while others, the caterpillar-like forms (Raupenform), resulted from them by a retrograde process, and also

---

* The Embryology of Chrysopa, and its bearings on the Classification of the Neuroptera, "American Naturalist," vol. v. Sept., 1871.

† "It is my opinion that the 'incomplete metamorphosis' of the Orthoptera is the primitive one, *inherited* from the original parents of all insects, and the 'complete metamorphosis' of the Coleoptera, Diptera, etc., a subsequently acquired one." *Fuer Darwin*, English Trans., p. 121.

Pl. 2.

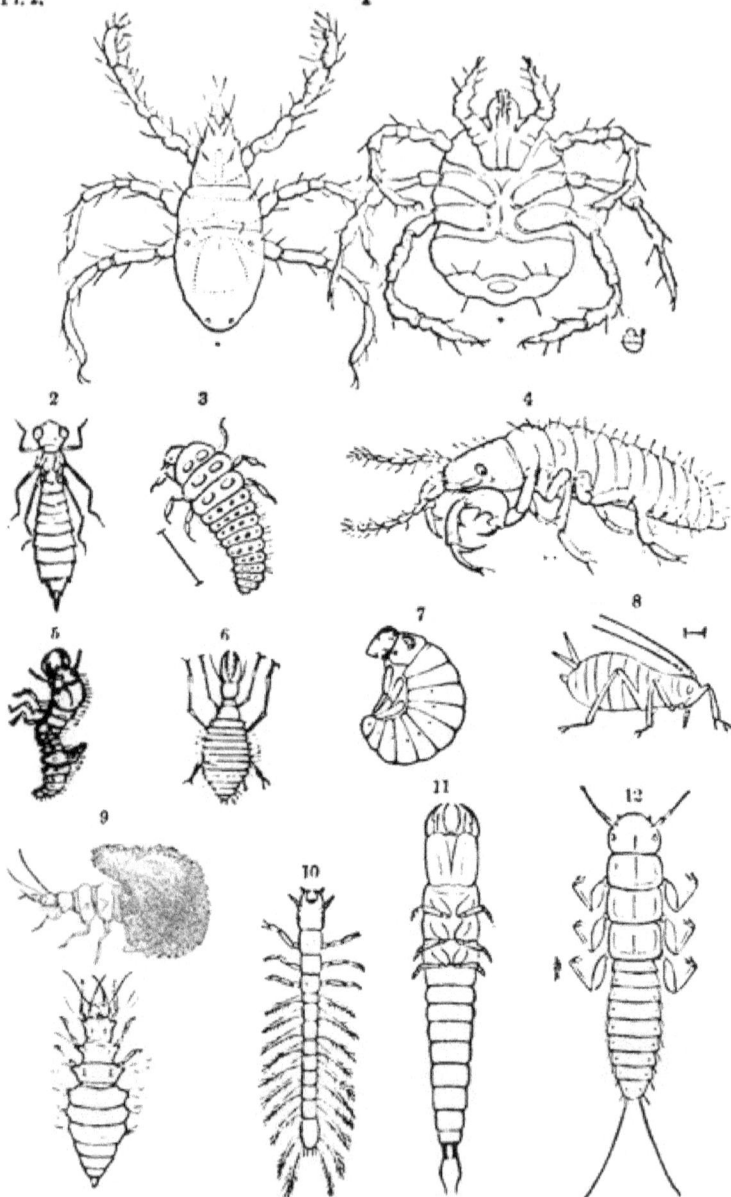

EXAMPLES OF LEPTIFORM LARVÆ.

EXPLANATION OF PLATE 2. Figure 1, different forms of Leptus; 2, Diplax; 3, Coccinella larva; 4, Cicada larva; 5, Cicindela larva; 6, Ant Lion; 7, Calligrapha larva; 8, Aphis larva; 9, Hemerobius larva; 10, Gyrinus larva; 11, Carabid larva; 12, Meloe larva.

Pl. 3.

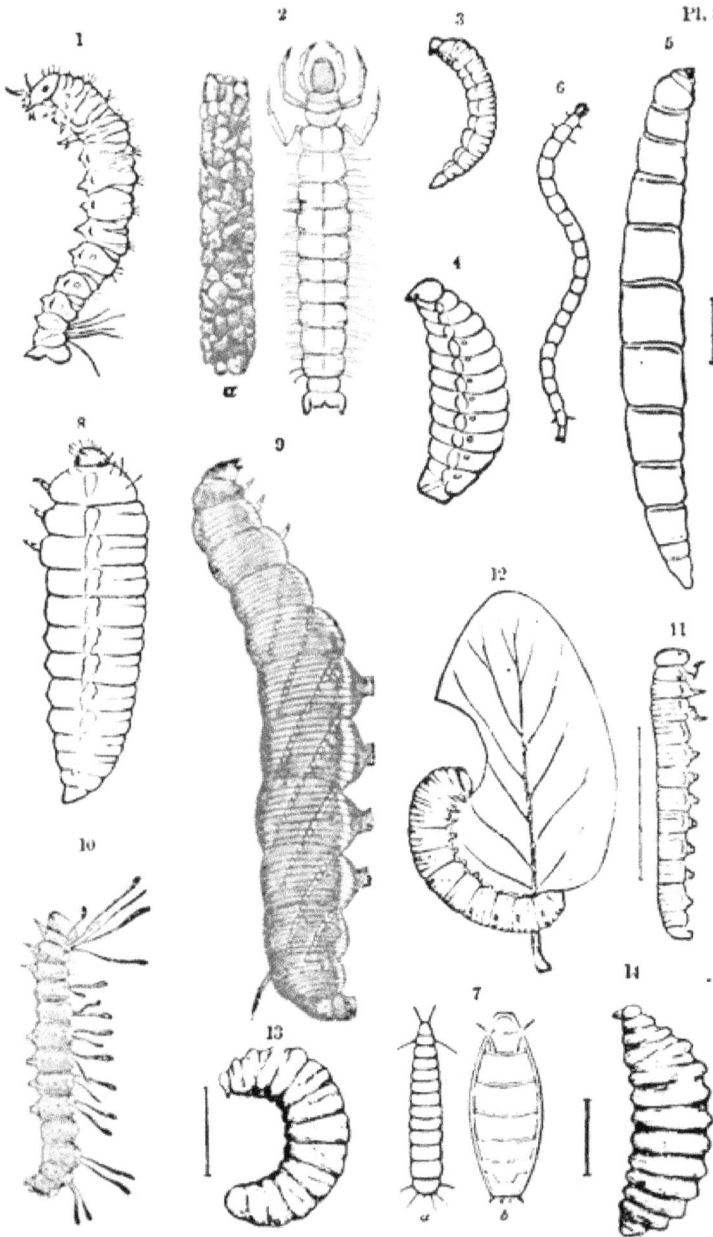

EXAMPLES OF ERUCIFORM LARVÆ.

EXPLANATION OF PLATE 3. Figure 1, Panorpa larva; 2, Phryganea larva; 3, Weevil larva; 4, third larva of Meloe; 5, Chionea larva; 6, Carpet Worm; 7, Phora larva; 8, Wheat Caterpillar; 9, Sphinx Caterpillar; 10, Acronycta? larva; 11, Saw Fly larva; 12, Abia Saw Fly larva; 13, Halictus larva; 14, Andrena larva.

the still lower maggot-like forms.  While on the one hand Campodea, with its abdominal feet, and the larva of Lithobius are related, so on the other the Lepismatidæ, which are very near the Blattariæ, are nearly related to the Myriopods, since their abdominal segments often bear appendages (Machilis).  The Campodea-form appears in most of the Pseudoneuroptera [Libellulids, Ephemerids, Perlids, Psocids and Termes], Orthoptera, Coleoptera, Neuroptera, perhaps modified in the Strepsiptera [Styleps and Xenos] and Coccidæ in their first stage of development, and indeed in many of these at their first moult."  Farther on he says, "A larger part of the most highly developed insects assume another larva-form, which appears not only as a later acquisition, through accommodation with certain definite relations, but also arises as such before our eyes.  The larvæ of butterflies and moths, of saw flies and Panorpæ, show the form most distinctly, and I call this the caterpillar form (Raupenform).  That this is not the primitive form, but one later acquired, we see in the beetles.  The larvæ of Meloë and Sitaris in their fully grown condition possess the caterpillar form, but the new-born larvæ of these genera show the Campodea form.  The last form is lost as soon as the larva begins its parasitic mode of life.  *  *  *  The larger part of the beetles, the Neuroptera in part, the bees and flies (the last with the most degraded maggot form) possess larvæ of this second form."  He considers that the caterpillar form is a degraded Campodea form, the result of its stationary life in plants or in wood.

204. Tipula Larva.

For reasons which we will not pause here to discuss, we have always regarded the cruciform type of larva as the highest.  That it is the result of degradation from the Leptus or Campodea form, we should be unwilling to admit, though the maggots of flies have perhaps retrograded from such forms as the larvæ of the mosquitoes and crane flies (Tipulids, Fig. 204).

That the cylindrical form of the bee grub and caterpillar is the result of modification through descent is evident in the caterpillar-like form of the immature Caddis fly (Pl. 3, fig. 2).  Here the fundamental characters of the larva are those of the Corydalus and Sialis and Panorpa, types of closely allied groups.  The features that remind us of caterpillars are superadded, evidently the result of the peculiar tube-inhabiting habits of

the young Caddis fly.   In like manner the caterpillar-form is
probably the result of the leaf-eating life of a primitive Leptiform larva.   In like manner the soft-bodied maggot of the
weevil is evidently the result of its living habitually in cavities
in nuts and fruits.   Did the soft, baggy female Stylops live
exposed, like its allies in other families, to an out-of-doors life,
its skin would inevitably become hard and chitinous.   In these
and multitudes of other cases the adaptation of the form of the
insect to its mode of life is one of cause and effect, and not a
bit less wonderful after we know what induced the change of
form.

Having endeavored to show that the caterpillar is a later
production than the young, wingless cockroach, with which
geological facts harmonize, we have next to account for the
origin of a metamorphosis in insects.   Here it is necessary to
disabuse the reader's mind of the prevalent belief that the
terms larva, pupa and imago are fixed and absolute.   If we
examine at a certain season the nest of a humble bee, we shall
find the occupants in every stage of growth from the egg to the
pupa, and even to the perfectly formed bee ready to break out of
its larval cell.   So slight are the differences between the different stages that it is difficult to say where the larval stage ends
and the pupa begins, so also where the pupal state ends and the
imago begins.   The following figures (205-208) will show four
of the most characteristic stages of growth, but it should be
remembered that there are intermediate stages between.   Now
we have noticed similar stages in the growth of a moth, though
a portion of them are concealed beneath the hard, dense chrysalis skin.   The external differences between the larval and pupal
states are fixed for a large part of the year in most butterflies
and moths, though even in this respect there is every possible
variation, some moths or butterflies passing through their transformations in a few weeks, others requiring several months,
while still others take a year, the majority of the moths living
under ground in the pupa state for eight or nine months.   The
stages of metamorphosis in the Diptera are no more suddenly
acquired than in the bee or butterfly.   In all these insects the
rudiments of the wings, legs, and even of the ovipositor of the
adult exist in the young larva.   We have found somewhat similar intermediate stages in the metamorphoses of the beetles.
The insects we have mentioned are those with a "complete

metamorphosis." We have seen that even in them the term "complete" is a relative and not absolute expression, and that the terms larva and pupa are convenient designations for states varying in duration, and assumed to fulfil certain ends of existence, and even then dependent on length of seasons, variation in climate, and even on the locality. When we descend to the insects with an "incomplete" metamorphosis, as in the May fly,

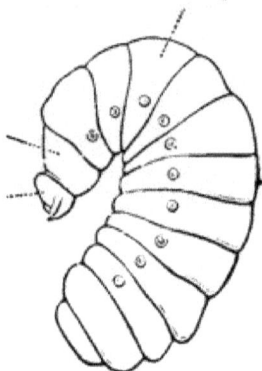

205. Larva.

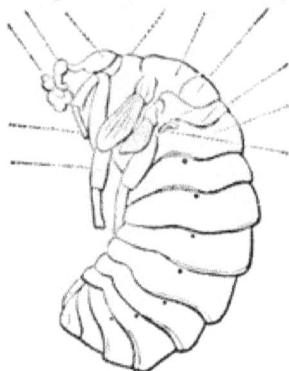

206. Semi-pupa.

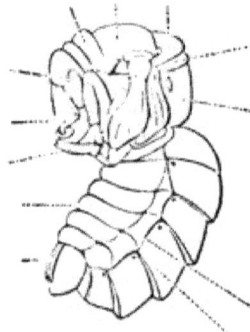

207. Advanced Semi-pupa.

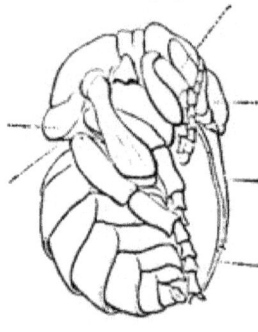

208. Pupa.

EARLY STAGES OF THE HUMBLE BEE.

we find that, as in the case of Chloëon, Sir John Lubbock has described twenty-one stages of existence, and let him who can say where the larval ends and the pupal or imaginal stages begin. So in a stronger sense with the grasshopper and cockroach. The adult state in these insects is attained after a number of moults of the skin, during each of which the insect gradually draws nearer to the final winged form. But even the so-called pupæ, or half winged individuals known not to be adult, in some cases feel the sexual impulse, while a number of species in each of the families represented by these two insects never acquire wings.

Still how did the perfect metamorphosis arise? We can only answer this indirectly by pointing to the Panorpa and Caddis flies, with their nearly perfect metamorphosis, though more nearly allied otherwise to those Neuroptera with an incomplete metamorphosis, as the lace-winged fly, than the insects of any other suborder. If, among a group of insects such as the Neuroptera, we find different families with all grades of perfection in metamorphosis, it is possible that larger and higher groups may exist in which these modes of metamorphosis may be fixed and characteristic of each. Had we more space for the exposition of many known facts, the sceptic might perceive that by observing how arbitrary and dependent on the habits of the insects are the metamorphoses of some groups, the fixed modes of other and more general groups may be seen to be probably due to biological causes, or in other words have been acquired through changes of habits or of the temperature of the seasons and of climates. Many facts crowd upon us, which might serve as illustrations and proofs of the position we have taken. For instance, though we have in tropics rainy and dry seasons when, in the latter, insects remain quiescent in the chrysalis state as in the temperate and frigid zones, yet did not the change from the earlier ages of the globe, when the temperature of the earth was nearly the same the world over, to the times of the present distribution of heat and cold in zones, possibly have its influence on the metamorphoses of insects and other animals? It is a fact that the remains of those insects with a complete metamorphosis (the bees, butterflies and moths, flies and beetles) abound most in the later deposits, while those with an incomplete metamorphosis are fewer in number and the earliest to appear. Again, certain groups of insects are not found in the polar regions. Their absence is evidently due to the adverse climatic conditions of those regions. The development of the same groups is striking in the tropics, where the sum of environing conditions all tend to favor the multiplication of insect forms.

It should be observed that some insects, as the grasshopper, for example, as Müller says, "quit the egg in a form which is dis-

16

tinguished from that of the adult insect almost solely by the want of wings," while the freshly hatched young of the bee, we may add, is farthest from the form of the adult. It is evident that in the young grasshoppers, the metamorphoses have been passed through, so to speak, in the egg, while the bee larva is almost embryonic in its build. The helpless young maggot of the wasp, which is fed solely by the parent, may be compared to the human infant, while the lusty young grasshopper, which immediately on hatching takes to the grass or clover field with all the enthusiasm of a duckling to its native pond, may be likened to that young feathered mariner. The lowest animals, as a rule, are at birth most like the adult. So with the earliest known crustacea. The king crabs, and in all probability the primeval trilobites, passed through their metamorphoses chiefly in the egg. So in the ancient Nebaliads (Peltocaris, Discinocaris and Ceratiocaris), if we may follow the analogy of the recent Nebalia, the young probably closely resembled the adult,

209. Jaws of Ant Lion.

while the living crabs and shrimps usually pass through the most marked metamorphoses. Among the worms, the highest, and perhaps the most recent forms, pass through the most remarkable metamorphoses.

Another puzzle for the evolutionist to solve is how to account for the change from the caterpillar with its powerful jaws, to the butterfly with its sucking or haustellate mouth-parts. We shall best approach the solution of this difficult problem by a study of a wide range of facts, but a few of which can be here noticed. The older entomologists divided insects into haustellate or suctorial, and mandibulate or biting insects, the butterfly being an example of one, and the beetle serving to illustrate the other category. But we shall find in studying the different groups that these are relative and not absolute terms. We find mandibulate insects with enormous jaws, like the Dytiscus, or Chrysopa larva or ant lion, perforated, as in the former, or enclosing, as in the latter two insects, the maxillae (b), which slide backward and forward within the hollowed mandibles (a, Fig. 209, jaws of the ant lion), along which the blood of their victims flows. They suck the blood, and do not tear the flesh of their prey. The enormous mandibles of the adult Corydalus are too large for use and, as Walsh observed, are converted in the male into simple clasping organs. And to omit a number of instances, in the suctorial Hemiptera or bugs we have different grades of structure in the mouth-parts. In the biting lice (Mallophaga) the mouth is mandibulate, in the Thrips it is mandibulate, the jaws being free, and the maxillæ bearing palpi, while the Pediculi are suctorial, and the true bugs are eminently so. But in the bed bug it is easy to see that the beak is made up of the two pairs of jaws, which are simply elongated and adapted for piercing and sucking. Among the so-called haustellate insects the mouth-parts vary so much in different groups, and such different organs separately or combined perform the function of sucking, that the term haustellate loses its significance and even misleads the student. For example, in the house fly the tongue (Fig. 210 *l*,

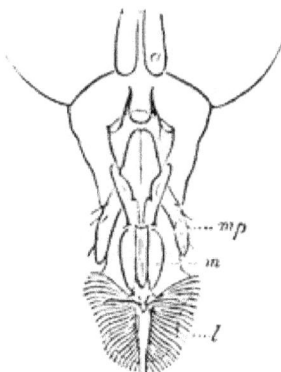

210. Mouth-parts of the House fly.

the mandibles, *m*, and maxillæ, *mp*, are useless), a fleshy prolongation of the labium or second maxillæ, is the sucker, while the mandibles and maxillæ are used as lancets by the horse fly (Fig. 211, *m*, mandibles, *mx*, maxillæ). The maxillæ in the butterfly are united to form the sucking tube, while in the bee the

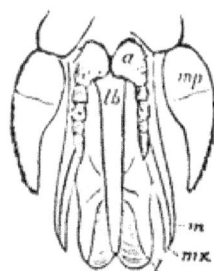

211. Mouth-parts of Horse fly.

end of the labium (Fig. 212) is specially adapted for lapping, not sucking, the nectar of flowers. But even in the butterfly, or more especially the moth, there is a good deal of misapprehension about the structure of the so-called "tongue." The mouth-parts of the caterpillar exist in the moth. The mandibles of the caterpillar occur in the head of the moth as two small tubercles (Fig. 213, *m*). They are aborted in the adult. While the maxillæ are as a rule greatly developed in the moth, in the caterpillar they are minute and almost useless. The labium or second maxillæ, so large in the moth, serves simply as a spinneret in the caterpillar. But we find a great amount of variation in the tongue or sucker of moths, and in the silk moths the maxillæ are rudimentary, and there is no tongue, these organs being but little more developed than in the caterpillar. Figure 213, B, shows the minute blade-like maxilla of the magnificent Luna moth, an approximation to the originally blade-like form in beetles and Neuroptera. The maxillæ in this insect are minute, rudimentary, and of no service to the creature, which does not take food. In other moths of the same family we have found the maxillæ longer, and touching at their tips, though too widely separate at base to form a sucking tube, while in others the maxillæ are curved, and meet to form a true tube.

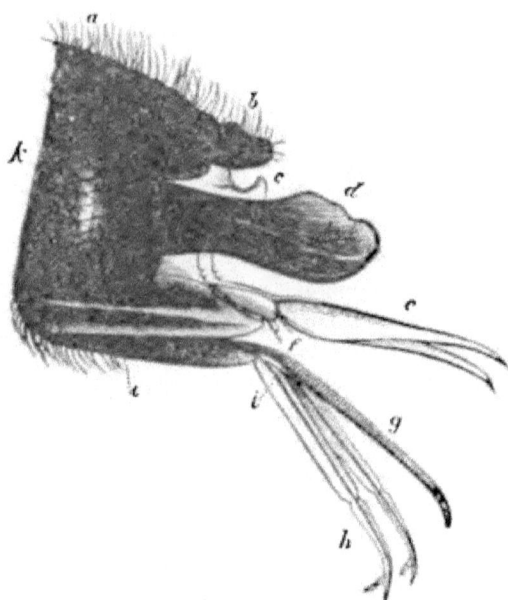

212. Head of Humble bee.

In the Cecropia moth it is difficult to trace the rudiments of the maxillæ at all, and thus we have in the whole range of the moths, every gradation from the wholly aborted

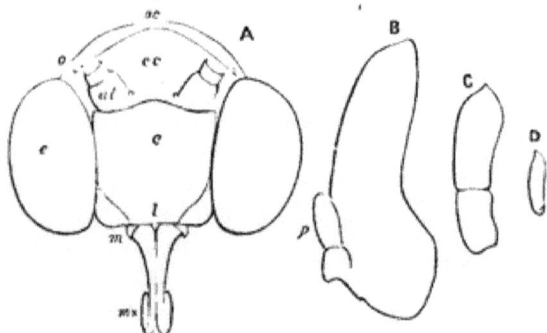

213. Mouth-parts of Moths.

maxillæ of the Platysamia Cecropia, to those of Macrosila cluentius of Madagascar, which form a tongue, according to Mr. Wallace, nine and a quarter inches in length, probably to enable their owner to probe the deep nectaries of certain orchids. These changes in form and size are certainly correlated with important differences in habits, and the evolutionist can as rightly say that the structural changes were induced by use and disuse and change of habits and the environment of the animal, as on the other hand the advocate of special creation claims that the two are simply correlated, and that is all we know about it.

Another set of organs, placed on quite another region of the body, unite to form the sting of the bee, or its equivalent the ovipositor of other hymenopterous insects, such as the Ichneumon fly (Fig. 214), the "saw" of the saw fly, and the augur of the Cicada. These are all formed on the same plan, arising early in the larval stage as three pairs of little tubercles, which ultimately form long blades, the innermost constituting the true ovipositor. We have found that one pair of these organs forms the "spring" of the Podura, and that in these insects it is three jointed, and thus is morphologically a pair of legs soldered together at their base. We would venture to regard the ovipositor of insects as probably representing three pairs of abdominal legs, comparable with those of the Myriopods, and even, as

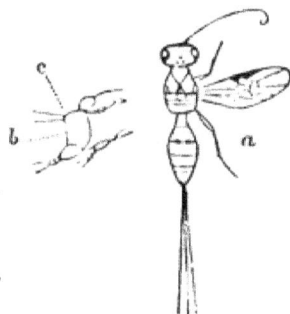

214. Ichneumon Fly.

we have suggested in another place, the three pairs of jointed spinnerets of spiders. Thus the ovipositor of the bee has a history, and is not apparently a special creation, but a structure gradually developed to subserve the use of a defensive organ.

So the organs of special sense in insects are in most cases simply altered hairs. The hairs themselves are modified epithelial cells. The eyes of insects, simple and compound, are at first simply epithelial cells, modified for a special purpose, and even the egg is but a modified epithelial cell attached to the walls of the ovary, which in turn is morphologically but a gland. Thus Nature deals in simples, and with her units of structure elaborates as her crowning work a temple in which the mind of man, formed in the image of God, may dwell. Her results are not the less marvellous because we are beginning to dimly trace the process by which they arise. It should not lessen our awe and reverence for Deity, if with minds made to adore, we also essay to trace the movements of His hand in the origin of the forms of life.

Some writers of the evolution school are strenuous in the belief that the evolution hypothesis overthrows the idea of archetypes, and plans of structure. But a true genealogy of animals and plants represents a natural system, and the types of animals, be they four, as Cuvier taught, or five, or more, are recognized by naturalists through the study of dry, hard, anatomical facts. Accepting, then, the type of articulates as founded in nature from the similar modes of development and points of structure perceived between the worms and the crustacea on the one hand, and the worms and insects on the other, have we not a strong genetic bond uniting these three great groups into one grand subkingdom, and can we not in imagination perceive the successive steps by which the Creator, acting through the laws of evolution, has built up the great articulate division of the animal kingdom?

# CHAPTER XIV.

In this calendar I propose to especially notice the injurious insects. References to the times of their appearance must be necessarily vague, and apply only, in a very general way, to the Northern States. Insects appear in Texas about six weeks earlier than in Virginia, in the Middle States six weeks earlier than in northern New England and the North-western States, and in New England about six weeks earlier than in Labrador. The time of the appearance of insects corresponds to the time of the flowering or leafing out of certain trees and herbs; for instance, the larvæ of the American Tent caterpillar and of the Canker worm hatch just as the apple tree begins to leaf out; a little later the Plant lice appear, to feast on the tender leaves; and when, during the first week in June, our forests and orchards are fully leafed out, hosts of insects are marshalled to ravage and devour their foliage.

## The Insects of Early Spring.

In April the gardener should scrape and wash thoroughly all his fruit trees, so as to rub off the eggs of the bark lice which hatch out early in May. Many injurious caterpillars and insects of all kinds winter under loose pieces of bark, or under matting and straw at the base of the trees. Search should also be made for the eggs of the Canker worm and the American Tent caterpillar, which last are laid in bunches half an inch long on the terminal shoots of many of our fruit trees. A little labor spent in this way will save many dollars' worth of fruit. The "castings" of the Apple Tree Borer (Saperda bivittata) should be looked for at the base of the tree, and its ravages be promptly

188     INSECT CALENDAR.

arrested. Its presence can also be detected, it is said, by the
dark appearance of the bark, where the grub is at work: cut in
and pull out the young grub. It is the best time of the year to
catch and kill this pest. Cylindrical bark borers, which are little round, black, weevil-like beetles, often causing "fire-blight"
in pears, etc., are now flying about fruit trees to lay their eggs;

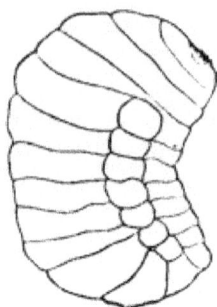

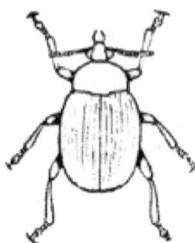

215. Pea Weevil and Maggot.

and many other weevils
and boring beetles, espe-
cially the Pea weevil (Bru-
chus pisi, Fig. 215), the
Pine weevil (Pissodes strobi, Fig. 216), and Hylobius
pales and Hylurgus tere-
brans, also infesting the
pine, now abound, and the
collector can obtain many
specimens not met with at
other times.

The housewife must now guard against the intrusion of
Clothes moths (Tinea), while many other species of minute
moths (Tineids) and of Leaf-rollers (Tortricidæ) will be flying
about orchards and gardens just as the buds are beginning to
unfold; especially the Coddling moth (Carpocapsa pomonella).
On warm days myriads of these and other insects may be seen
filling the air; it is the busiest time of their lives, as all are on
errands of love to their
kind, but of mischief to
the agriculturist.

When the May Flower
—"O commendable flowre
and most in minde"—
blooms, and the willows
hang out their golden cat-
kins, we shall hear the
hum of the wild bee, and

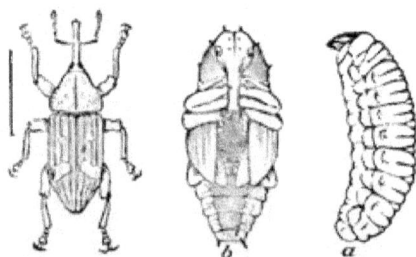

216. Pine Weevil and Young.

the insect hunter will reap a rich harvest of rarities. Seek now
on the abdomen of various wild bees, such as Andrena, for that
most eccentric of all our insects, the Stylops Childreni. The
curious larvæ of the Oil beetle may be found abundantly on the
bodies of various species of Bombus, Andrena and Halictus, with
their heads plunged in between the segments of the bee's body.

The beautiful moth, Adela, with its immensely long antennæ, may be seen, with other smaller moths, feeding on the blossoms of the willow. The Ants wake from their winter's sleep and throw up their hillocks, and the "thriving pismire" issues from his vaulted galleries constructed in some decaying log or stump, while the Angle worms emu-

217. The Comma Butterfly.

late their six-footed neighbors. During the mild days of March, ere the snow has melted away—

"The dandy Butterfly,
All exqui-itely dre-t,"

will visit our gardens. Such are various kinds of Vanessa and Grapta (Fig. 217, G. c-argenteum*). The beautiful Brephos in-fans flies before the snow disappears.

"The Gnat, old back-bent fellow,
In frugal frieze coat drest,"

will celebrate the coming of Spring, with his choral dance. Such is Trichocera hyemalis, which may be seen in multitudes towards twilight on mild even-ings. Many flies are now on the wing, such as Tachina (Fig. 218) and its allies; the four spotted Mosquito, Anopheles quadri-maculatus, and the delicate spe-cies of Chironomus, whose males have such beautifully feathered antennæ, assemble in swarms. Now is the time for the collector to turn up stones and sticks by the river's side

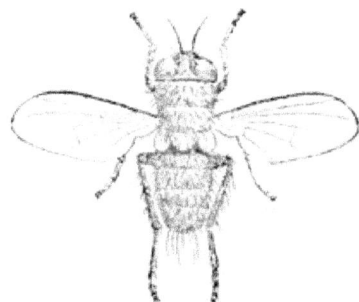

218. Tachina.

and in grassy damp pastures, for Ground beetles (Carabidæ), and to frequent sunny paths for the gay Cicindela and the Bom-

*The right side represents the under side of the wings.

bylius fly, or fish in brooks and pools for water beetles and various larvæ of Neuroptera and Diptera; while many flies and beetles are attracted to freshly cut maples or birches running with sap; indeed, many insects, rarely found elsewhere, assemble in quantities about the stumps of these trees, from which the sap oozes in March and April.

In April the injurious insects in the Northern States have scarcely begun their work of destruction, as the buds do not unfold before the first of May. We give an account, however, of some of the beneficial insects which are now to be found in grass-lands and in gardens. The farmer should know his true insect friends as well as his insect foes. We introduce to our readers a large family of ground-beetles (Carabidæ, from Cara-

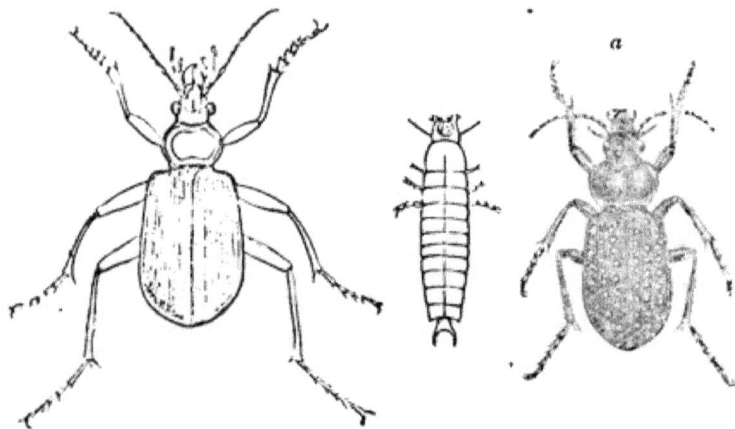

219. Calosoma scrutator.        220. Calosoma calidum and Larva.

bus, the name of the typical genus) which prey on those insects largely injurious to crops. A study of the figures will familiarize our readers with the principal forms. They are dark-colored, brown or black, with metallic hues, and are seen in spring and throughout the summer, running in grass, or lurking under stones and sticks in damp places, whence they sally forth to hunt by night, when many vegetable-eating insects are most active.

The larvæ are found in much the same situations as the mature beetles. They are elongate, oblong, and rather broad, the terminal ring of the body being armed with two horny hooks, and having a single fleshy leg beneath; and are usually black in color. The larva of Calosoma (C. calidum, Fig. 220; a, the beetle; and Fig. 219, C. scrutator) ascends trees to feed on caterpillars, such as the Canker worm. When about to trans-' form to the pupa state, it forms a rude cocoon in the earth. The beetle lies in wait for its prey in shallow pits excavated in pastures. We once saw it fiercely attack a May beetle (Lachnosterna fusca) nearly twice its size; it tore open the hard sides of its clumsy and helpless victim with tiger-like ferocity. Carabus (Fig. 221, C. serratus Say, and pupa of Carabus auronitens of Europe, after West-wood) is a closely allied form, with very similar habits.

A much smaller form is the curious Bombardier beetle, Brachinus (Fig. 222. B. fumans), with its nar-

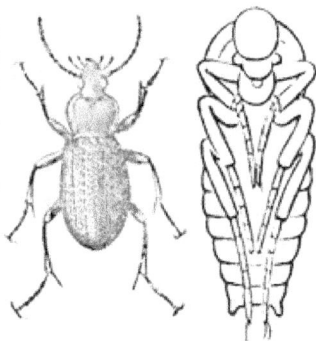

221. Carabus and Pupa.

row head and heart-shaped prothorax. It is remarkable for discharging with quite an explosion from the end of its body a pungent fluid, probably as a protection against its enemies. An allied genus is Casnonia (Fig. 223, C. Pensylvanica), which

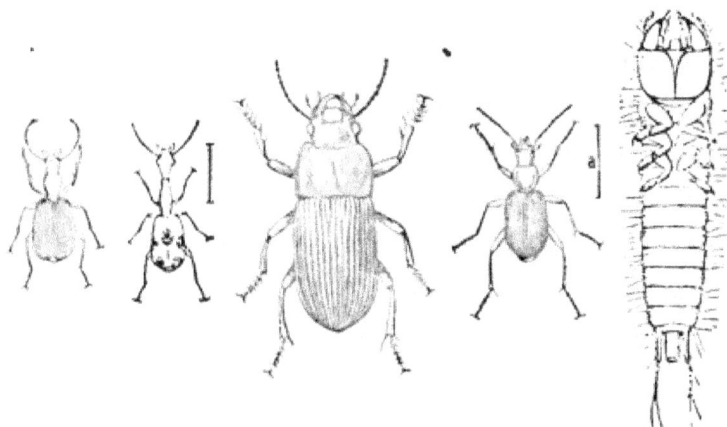

222.        223.           224.            225.       226. Carabid
Brachinus.  Casnonia.     Pangus.        Agonum.    Larva.

has a long neck and spotted wing covers. Figure 224, Pangus caliginosus, and figure 225, Agonum cupripenne, represent two common forms. The former is black, while the latter is a pretty insect, greenish, with purplish-red wing-covers, and black legs.

Figure 226, enlarged about three times, represents a singular larva found by Mr. J. H. Emerton under a stone early in spring. Dr. LeConte, to whom we sent a figure, supposes that it may possibly be a larva of Harpalus, or Pangus caliginosus. It is evidently a young Carabid. The under side is represented.

## The Insects of May.

During this month there is great activity among the insects. As the flowers bloom and the leaves appear, multitudes wake from their long winter sleep, and during this month pass through the remainder of their transformations, and prepare for the summer campaign. Most insects hibernate in the chrysalis or pupa state, while many winter in the caterpillar or larva state, such as the larvæ of several Noctuidæ and the "yellow-bear," and other caterpillars of Arctia and its allies. Other insects hibernate in the adult or imago form, either as beetles, butterflies or certain species of bees.

It is well known that the Queen Humble bee winters under the moss, or in her old nest. During the present month her rovings seem to have a more definite object, and she seeks some deserted mouse's nest, or hollow in a tree or stump, and there stows away her pellets of pollen, containing two or three eggs apiece, which, late in the summer, are to form the nucleus of a well-appointed colony. The Carpenter bees (Ceratina and Xylocopa, the latter of which is found in abundance south of New England) are busy in refitting and tunnelling the hollows of the grape; while the Ceratina hollows out the stem of the elder, or blackberry. This little upholsterer bee carpets her honey-tight apartment, storing it with food for her young, and later in the season, in June, several of these cartridge-like cells, whose silken walls resemble the finest and most delicate parchment, may be found in the hollow stems of these plants. The Mason bee (Osmia) places her nest in a more exposed site, building her earthen cells of pellets of moistened mud, either situated under a stone, or in some more sheltered place; for instance, in a deserted oak-gall, ranging half a dozen of them side by side along the vault of this strange domicile. Meanwhile their more lowly relatives, the Andrena and Halictus bees, are engaged in tunnelling the side of some sunny bank or path, running long galleries underground, sometimes for a foot or more, at the farthest end of which are to be found, in summer, little earthen

urn-like cells, in which the grubs live upon the pollen stored
up for them in little balls of the size of a pea. Later in the
month, the Gall flies (Cynips),
those physiological puzzles,
sting the leaves of our oaks
of different species, giving rise
to the strange excrescences and
manifold deformities which
deface the stems and leaves
of our most beautiful forest
trees.

227. Chrysophanus Thoe.*

When the Kalmia, Rhodora, and wild cherries are in bloom,
many of our most beautiful butterflies appear; such are the

228. Argynnis Aphrodite.

different species of Chrysophanus (Fig. 227), Lycæna, Thecla and
Argynnis (Fig. 228). At this time we have found the rare larva
of Melitæa Phaeton
(Fig. 229) clothed in
the richest red and
velvety black, feed-
ing daintily upon
the hazel nut, and
tender leaves of the
golden rod. In June,
it changes to the
chrysalis state, and

229. Melitæa Phaeton.

early in July the butterfly rises from the cold, damp bogs, where

*The lower side of the wings is figured on the right side of this and Figs. 228 and 229.

17

we have oftenest found it, clad in its rich dress of velvety black and red.

Later still, when the lilac blooms, and farther south the broad-leaved Kalmia, the gaily-colored Humming Bird moth (Sesia) visits the flowers in company with the Swallow-tail butterfly (Papilio Turnus). At twilight, the Hawk moth (Sphinx) darts noiselessly through our gardens, as soon as the honeysuckles, pinks and lilies are in blossom.

230, D. 12-punctata. 231. Diabrotica vittata.

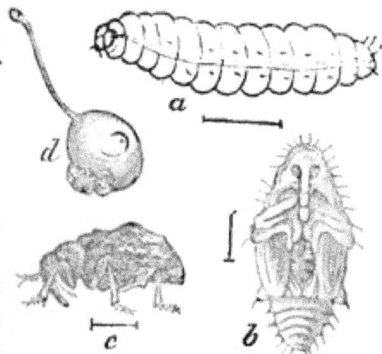

Among the flies, mosquitoes now appear, though they have not yet, perhaps, strayed far from their native swamps and fens; and their mammoth allies, the Daddy-long-legs (Tipula), rise from the fields and mould of our gardens in great numbers.

Of the beetles, those which feed on leaves now become specially active. The Squash beetle (Diabrotica vittata, Fig. 231, and Fig. 230, D. 12-punctata) now attacks the squash plants before they are fairly up; and the Plum weevil (Conotrachelus nenuphar, Fig. 232) will sting the newly formed fruit, late in the month, or early in June.

Fig. 232. Plum Weevil and Young.

Many other weevils now abound, stinging the seeds and fruit, and depositing their eggs just under the skin. So immense are the numbers of insects which fill the air and enliven the fields and woodlands just as summer comes in, that a bare enumeration of them would overcrowd our pages, and tire the reader.

A word, however, about our water insects. Late in the month the May fly (Ephemera, Fig. 233) appears, often rising in immense numbers, from the surface of pools and sluggish brooks.

233. May Fly.

In Europe, whole clouds of these delicate forms, with their thin white wings, have been

known to fall like snow upon the ground, when the peasants gather them up in heaps to enrich their gardens and farms.

The Case worms, or Caddis flies (Fig. 234), begin now to leave their portable houses, formed of pieces of leaves, or sticks and fine gravel, or even of shells, as in an European species, and fly over the water, resting on the overhanging trees.

A few busy Mosquito Hawks, or Dragon flies (Libellula), herald the coming of the summer brood of these indefatigable friends of the agriculturist. During their whole life below the waters, these entomological Herods have slain and sucked the blood of myriads of infant mosquitoes and other insects; and now in their new world above the waters, with still more intensified powers of doing mischief, happily, however, to flies mostly obnoxious to man, they riot in bloodshed and carnage.

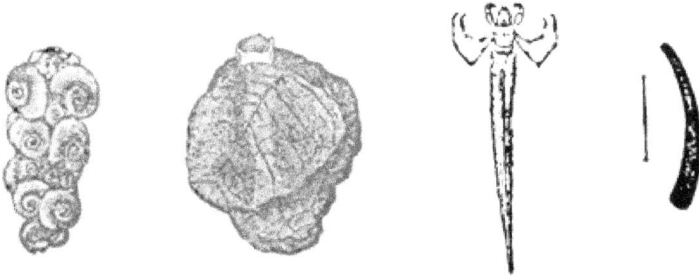

234. Different Forms of Case Worms.

This is the season to stock the fresh-water aquarium. Go to the nearest brook, gather a sprig or two of the water cress, which spreads so rapidly, a root of the eel grass, and plant them in a glass dish or deep jar. Pour in your water, let the sand and sediment settle, and then put in a few Tadpoles, a Newt (Salamander), Snails (Limnæa, Planorbis and Valvata), Caddis flies and Water beetles, together with the gatherings from a thicket of eel grass, or other submerged plants, being rich in the young of various flies, Ephemeras, Dragon flies and Water fleas (Entomostraca, Fig. 235), which last are beautiful objects for the microscope, and in a few days the occupants will feel at home, and the aquarium will be swarming with life, affording amusement and occupation for many a dull hour, by day or at night, in watching the marvels of insect transformations, and plant-growth.

Among the injurious hymenoptera, which abound late in this

month, is the Rose Saw fly (Selandria rosæ, Fig. 236) and S.
cerasi. The eggs are then laid, and the last of June, or early in

July, the slug-like larvæ mature, and
the perfect insects fly in July. Various
Gall flies now lay their eggs in the
buds, leaves and stems of various kinds
of oaks, blackberries, blueberries and
other plants.

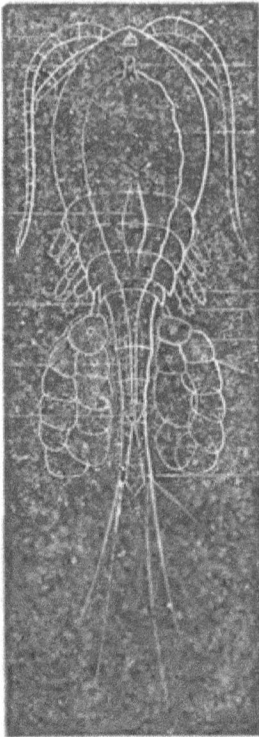

Dipterous Gall flies, are now laying
their eggs in cereals. The Hessian
fly (Cecidomyia destructor) has two
broods, the fly appearing both in
spring and autumn. The fly lays
twenty or thirty eggs in a crease in
the leaf of the young plant. In about
four days, in warm weather, they
hatch, and the pale-red larvæ "crawl
down the leaf, work-
ing their way in be-
tween it and the main
stalk, passing down-
ward till they come
to a joint, just above
which they remain, a
little below the sur-
face of the ground,
with the head tow-
ards the root of the

235. Water Flea.

plant. Here they imbibe the sap by suction
alone, and, by the simple pressure of their
bodies become imbedded in the side of the
stem. Two or three larvæ thus imbedded
serve to weaken the plant and cause it to wither
and die. The second brood of larvæ remains
through the winter in the flax-seed, or puparium. By turning the stubble with the plough
in the autumn and early spring, its imago may
be destroyed, and thus its ravages may be
checked. (Figure 237 represents the female,
which is about one-third as large as a mosquito : a, the larva;
b, the pupa; and c represents the joint near the ground where

236. Selandria rosæ.

the maggots live.)   The same may be said of the Wheat midge
(Cecidomyia tritici), which attacks the wheat in the ear, and
which transforms an inch deep beneath the surface.

Among the butterflies which appear this month are the Tur-
nip-butterfly (Pontia oleracea, Fig. 238,) which lays its eggs the
last of the month.   The
eggs hatch in a week or
ten days, and in about two
weeks the larva changes
to a chrysalis.   Thanaos
junevalis and T. Brizo fly
late in May.   The cater-
pillars live on the pea
and other papilionaceous
plants.   Thecla Auburni-
ana, T. Niphon, and other
species fly in dry, sunny
fields, some in April.   Ar-

238. Turnip Butterfly.

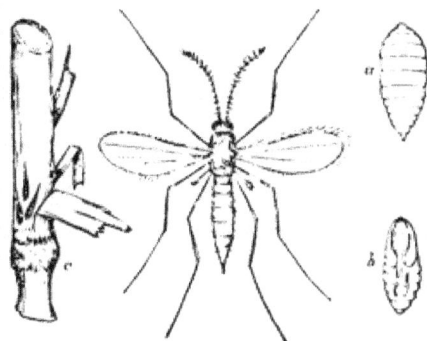
237. Hessian Fly.

gynnis Myrina flies from the last of May through June, and a
second brood appears in August and September.   Vanessa
J-album and V. interrogationis appear in May, and again in
August and September.   The caterpillars of the latter species
live on the elm, lime and hop-vine.   Grapta comma also feeds
on the hop.   Alypia 8-mac-
ulata (Fig. 49) flies at this
time, and in August its
larva feeds on the grape.
Sphinx gordius, S. 5-macu-
lata (Fig. 239) and other
Sphinges and Sesia (the
Clear-winged moth),
appear the last of May.
Arctia Arge, A. virgo, A.
phalerata and other spe-
cies fly from the last of
May through the summer.

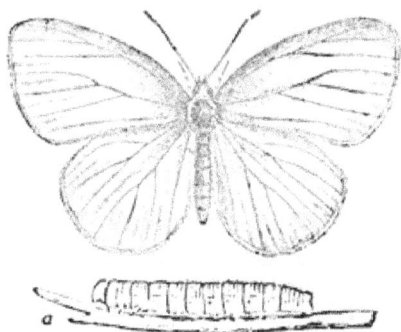
238. Turnip Butterfly.

Hyphantria textor, the Fall-weaver, is found in May or June.
The moth of the Salt-marsh caterpillar appears at this time,
and various Cut worms (Agrotis, Fig. 240) abound, hiding in
the daytime under stones and sticks, etc., while various Tineids
and Tortrices, or Leaf-rolling caterpillars, begin to devour ten-

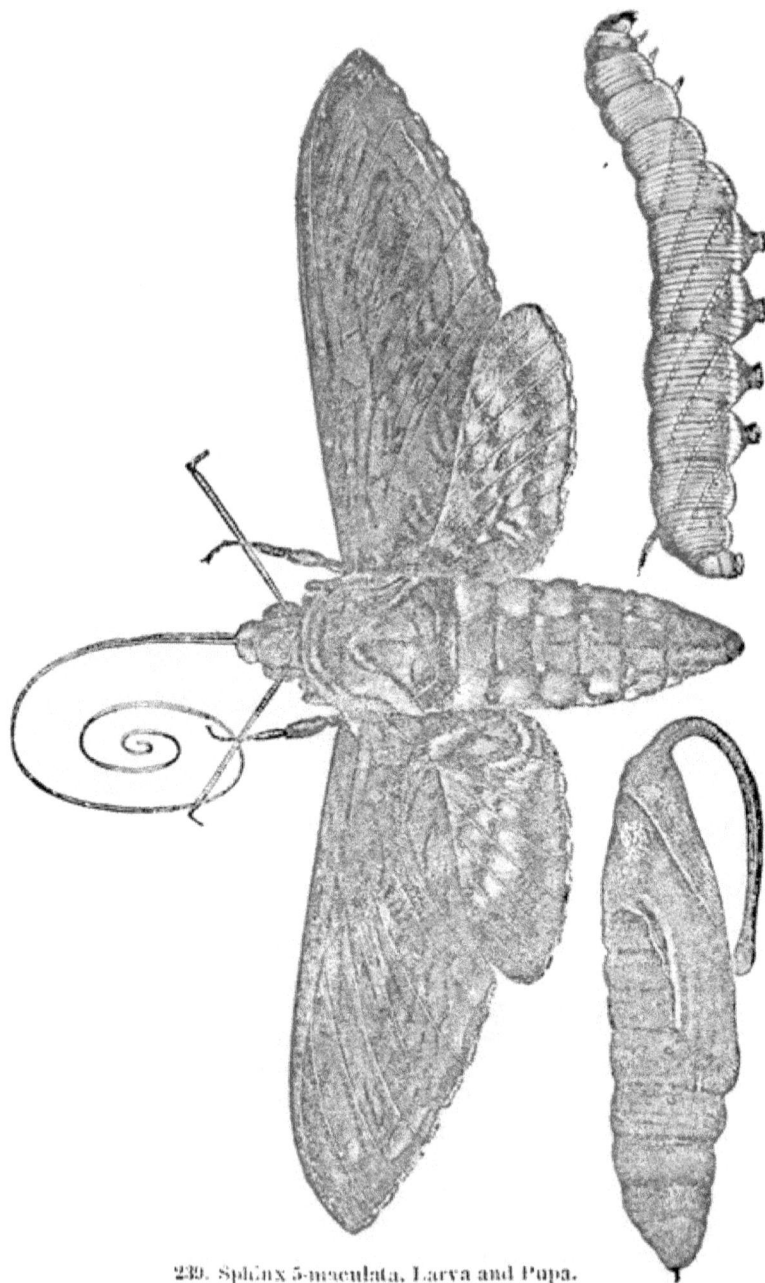

239. Sphinx 5-maculata, Larva and Pupa.

der leaves and buds and opening blossoms of flowers and fruit trees.

The White-pine weevil flies about in warm days. We have found its burrows winding irregularly over the inner surface of the bark and leading into the sap-wood. Each cell, in which it hibernates, in the middle of March, contains the yellowish white footless grub. Early in April it changes to a pupa, and a month after the beetle appears, and in a few days deposits its egg under the bark of old pine trees. It also oviposits in the terminal shoots of pine saplings, dwarfing and permanently deforming the tree. Associated with this weevil we have found the smaller, rounder, more cylindrical, whitish grubs of the

210. Cut Worm and Moth.

Hylurgus terebrans, which mines the inner layers of the bark, slightly grooving the sap-wood. Later in April it pupates, and its habits accord in general with those of Pissodes strobi. Another Pine weevil also abounds at this time, as well as Otiorhynchus picipes (Fig. 211), which injures beans, etc.

Cylindrical bark-borers, which are little, round, weevil-like beetles, are now flying about fruit trees, to lay their eggs in the bark. Associated with the Pissodes, we may find in April the galleries of Tomicus pini, branching out from a common centre. They are filled up with fine sawdust, and, according to Dr. Fitch, are notched in the sides "in which the eggs have been placed, where they would remain undisturbed by the beetle as it crawled backwards and forth

211. Garden Weevil.

through the gallery." These little beetles have not the long snouts of the weevils, hence they cannot bore through the outer bark, but enter into the burrows made the preceding year, and distribute the eggs along the sides (Fitch). Another Tomicus, more dangerous than the preceding, feeds exclusively

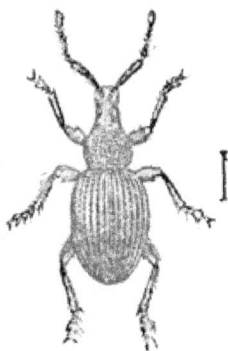

in the sap-wood, running solitary galleries for a distance of two
inches towards the centre of the tree. We figure Tomicus
xylographus Say (Fig. 242, enlarged). It is the most formidable enemy to the white pine in the North, and the yellow pine
in the South that we have. It also flies in May. Ptinus fur

242. Pine Weevil.                    243. Ptinus and Larva.

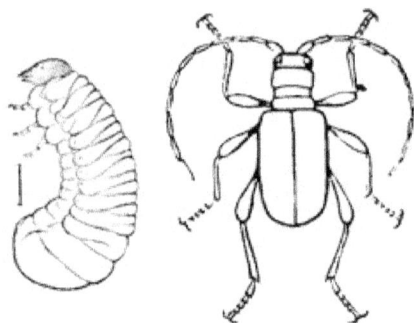

(Fig. 243. much enlarged) is now found in out-houses. and is
destructive to cloth, furs, etc., resembling the Larder-beetle
(Dermestes) in its habits. It is fourteen hundredths of an inch
in length.

### The Insects of June.

Early in the month the Parsnip butterfly (Papilio Asterias)
may be seen flying about, preparatory to laying its eggs for the
brood of caterpillars which appear in August. At the time of
the flowering of the raspberry and blackberry, the young larva
of Vanessa Antiopa, one of our most abundant butterflies, may
be found living socially on the leaves of the willow; while the
mature larva of another much smaller butterfly, the little Copper
skipper (Chrysophanus Americanus), so abundant at this time,
may sometimes be found on the clover. It is a short, oval,
greenish worm, with very short legs. The dun-colored skippers
(Hesperia) abound towards the middle of the month, darting
over the flowers of the blueberry and blackberry, in sunny
openings in the forests.

The family of Hawk moths (Sphinges) now appear in greater
abundance, hovering at twilight over flower-beds, and, during
this time, deposit their eggs on the leaves of various fruit-trees.
The American Tent caterpillar makes its cocoon, and assumes
the pupa state. The caterpillar passes several days within the

cocoon, in what may be called the semi-pupa state, during which period the chrysalis skin is forming beneath the contracted and loosened larva skin. We once experimented on a larva which had just completed its cocoon, to learn how much silk it could produce. On removing its cocoon it made another of the same thickness; but on destroying this second one it spun a third but frail web, scarcely concealing its form. A minute Ichneumon parasite, allied to Platygaster, lays its eggs within those of this moth, as we once detected one under a bunch of eggs, and afterwards reared a few from the same lot of eggs. A still more minute egg-parasite (Fig. 244) we have seen ovipositing in the early spring, in the eggs of the Canker-worm.

Among that beautiful family of moths, the Phalænidæ, comprising the Geometers, Loopers, or Span-worms, are two formidable foes to fruit growers. The habits of the Canker worm should be well known. With proper care and well-directed energy, we believe their attacks can be in a great measure prevented. The English sparrow, doves and other insectivorous birds, if there are any others that eat them, should be domesticated in order to reduce the number of these pests. More care than has yet been taken should be devoted to destroying the eggs laid in the autumn, and also the wingless females, as they crawl up the trees in the spring and autumn to lay their eggs. The evil is usually done before the farmer is well aware that the calamity has fallen upon him. As soon as, and even before the trees have fairly leafed out, they should be visited morning, noon and night, shaken and thoroughly examined and cleared of the caterpillars. By well-concerted action among agriculturists, who should form a Board of Destruction, numbering every man, woman and child on the farm, this fearful scourge may be abated by the simplest means, as the cholera or any epidemic disease can in a great measure be averted by taking proper sanitary precautions. The Canker worms hatch out during the early part of May, from eggs laid in the fall and spring, on the branches of various fruit-trees. Just as the buds unfold, the young caterpillars make little holes through the tender leaves, eating the pulpy portions, not touching the veins and midribs. When four weeks old they creep to the ground, or let themselves down by spinning a silken thread, and burrow from two to six inches in the soil, where

244. Canker worm Egg-parasite.

they change to chrysalids in a day or two, and in this state live
till late in the fall, or until the early spring, when they assume
the imago or moth form.  The sexes then unite, and the eggs
are deposited for the next generation.

The Canker worm is widely distributed, though its ravages
used to be confined mostly to the im-
mediate vicinity of Boston.  We have
seen specimens of the moth from Illinois.  Riley has found it in Missouri.

The Abraxas ribearia of Fitch
(Fig. 245, moth), the well-known
Currant worm, defoliates whole rows
of currant bushes.  This pretty cater-

245. Abraxas ribearia.

pillar may be easily known by its body being of a deep golden
color, spotted with black.  The bushes should be visited morning, noon and night, and thoroughly shaken (killing the caterpillars) and sprinkled with ashes.

Among multitudes of beetles (Coleoptera) injurious to the

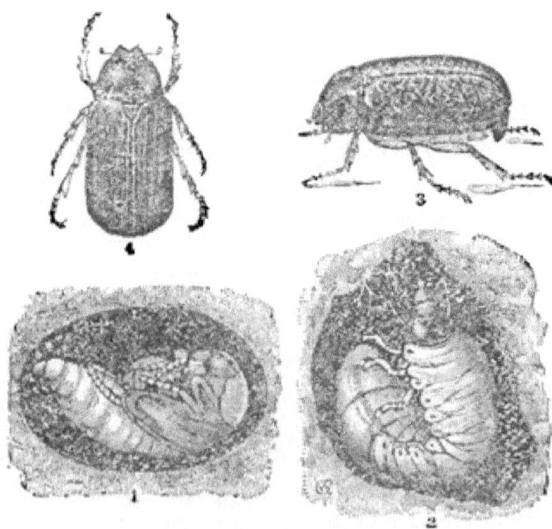

246. May Beetle and Young.

crops, are the May beetle (Lachnosterna fusca, Fig. 246), whose
larva, a large white grub, is injurious to the roots of grass and
to strawberry vines.  The Rose beetle appears about the time
of the blossoming of the rose.  The Fire-flies now show their

light during mild evenings, and on hot sultry days the shrill
rasping song of the male Cicada, for "they all have voiceless
wives," cuts the air. The Chinch-bug, that fell destroyer of our
wheat crops, appears, according to Harris, in the middle of the
month, and "may be seen in their various stages of growth on
all kinds of grain, on corn
and herds-grass during
the whole summer." So
widely spread is this in-
sect at present, that we
have even detected it in
August on the summit of
Mount Washington.

247. Pemphigus.

The Diptera, or two-winged flies, contain hosts of noxious
insects, such as the various Cecidomyians, or two-winged Gall
flies, which now sting the culms of the wheat and grasses, and
various grains, and leaves of trees, producing gall-like excrescences of varying form. Legions of these delicate minute flies
fill the air at twilight, hovering over wheat fields and
shrubbery. A strong north west wind, at such times,
is of incalculable value to the farmer. Moreover,
minute flies, allied to the house fly, such as Tephritis,
Oscinis, etc., now attack the young cereals, doing
immense injury to grain.

248. Apple
Bark Louse.

Millions of Aphides, or Plant lice, now infest our
shade and fruit trees, crowding every green leaf,
into which they insert their tiny beaks, sucking in
the sap, causing the leaves to curl up and wither.
They also attack the stems and even the roots of
plants, though these latter (Pemphigus, Fig. 247)
differ generically from the true Plant lice. Fruit
trees should be again washed and rubbed to kill off
the young Bark lice, of which the common apple
Bark louse (Aspidiotus conchiformis, Fig. 248),
whose oyster-shaped scales may be found in myriads
on neglected trees, is a too familiar example. Another pest of apple trees is the woolly Blight (Eri-
osoma lanigera). These insects secrete from the surface of the
body a downy, cottony substance which conceals the animal, and
when they are, as usual, grouped together on the trees, makes
them look like patches of mould. The natural insect enemies

of the Plant lice now abound; such are the Lady bugs (Coccinella, Fig. 249); the larva of the Syrphus fly (Fig. 76), which
devours immense quantities, and the larva of the Golden-eyed,
Lace-winged fly (Chrysopa, Fig. 256).

The last days of June are literally the heyday and jubilee of
insect life. The entomological world holds high carnival, though
in this country they are, perhaps, more given to mass-meetings
and caucuses. The earth, the air, and the
water teem with insect life. The insects
of mid-summer now appear. Among the
butterflies, the Wood Satyrus (Neonympha
Eurythris) skips in its low flight through
the pines. The larva of Grapta Progne
appears on the currants, and feeds beneath
the leaves on hot sunny days. The larva

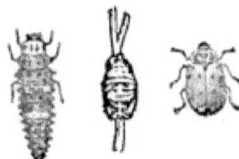

249. Coccinella and Young.

of Cynthia cardui may be found on the hollyhocks; the pupa
state lasts twelve days, the butterfly appearing in the middle
or last of July. The Hyphantria textor now lays its smooth,
spherical eggs in broad patches on the under side of the leaves
of the apple, which the caterpillar will ravage in August; and
its ally, the Halesidota caryæ, we have found ovipositing the
last week in the month on the leaves of the butternut. The
Squash bug, Coreus (Gonocerus) tristis (Fig. 250) is now very
abundant, gathering about the roots of the squash vines, often
in immense numbers, blackening the stems with
their dark, blackish-brown bodies. This insect
is easily distinguished from the yellow striped
Squash beetle previously mentioned, by its much
greater size, and its entirely different structure
and habits. It is a true bug (Hemipter, of which
the bed-bug is an example), piercing the leaves
and stalks, and drawing out the sap with its long
sucker.

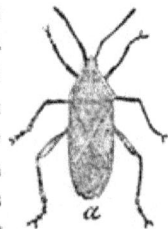

250. Squash Bug.

In June, also, we have found that beautiful butterfly, Militæa
Phaeton rising from the low, cold swamps. Its larva transforms
early in June or the last week in May, into a beautiful chrysalis.
The larva hibernates through the winter, and may be found
early in spring feeding on the leaves of the aster, the Viburnum
dentatum and hazel. It is black and deep orange-red, with
long, thick-set, black spines.

The Currant borer, Trochilium tipuliforme (Fig. 251), a beautiful, slender, agile, deep blue moth, with transparent wings, flies the last of the month about currant bushes, and its chrysalids may be found in May in the stems. Among moths, that of the American Tent caterpillar flies during the last of June and July, and its white cocoons can be detected under bark, and in sheltered parts of fences and out-houses.

Among others of the interesting group of Silk worms (Bombycidæ) are Lithosa, Crocota and allies, which fly in the day-time, and the different species of Arctia, and the white Arctians, Spilosoma, and Leucarctia, the parent of the Salt-marsh Caterpillar.

Many Leaf rollers, Tortrices, are rolling up leaves in various ways for their habitations, and to conceal them from too prying birds; and hosts of young Tineans are now mining leaves, and excavating the interior of seeds and various fruits. Grape-growers should guard against the attacks of a species of Tortrix (Penthina vitivorana) which rolls the leaves of the grape, and, according to Mr. M. C. Reed, of Hudson, Ohio, "in mid-summer deposits its eggs in the grape; a single egg in a grape. Its presence is soon indicated by a reddish color on that side of the yet green grape, and on opening it, the winding channel opened by the larva in the pulp is seen, and the minute worm, which is white, with a dark head, is found at the end of the

251. Currant Moth.

channel. It continues to feed upon the pulp of the fruit. and when it reaches the seeds, eats out their interior; and if the supply from one grape is extinguished before its growth is completed, it fastens this to an adjoining grape with a web, and burrows into it. It finally grows to about one-half of an inch in length, becomes brown, almost black, the head retaining its cinnamon color. When it leaves the grape it is very active, and has the power of letting itself down by a thread of silk. All my efforts to obtain the cocoons failed until I placed fresh grape leaves in the jar containing the grapes. The larvæ immediately betook themselves to these, and, cutting a curved line through the leaf thus ), sometimes two lines thus ( ). folded the edge or edges over, and in the fold assumed the chrysalis form. From specimens saved, I shall hope to obtain the perfect insect this season, and perhaps obtain information which will aid in checking its increase. Already it is so abundant that it is necessary to examine every branch of ripe grapes, and clip

18

out the infested berries before sending them to the table. A rapid increase in its numbers would interfere seriously with the cultivation of the grape in this locality."

The Rose beetle (Macrodactyla subspinosa) appears in great abundance. The various species of Buprestis are abundant; among them are the Peach-borer (Dicerca divaricata), which may be now found flying about peach and cherry trees; and Chrysobothris fulvogutta, and C. Harrisii, about white pines. A large weevil (Arrhenodes septentrionalis), which lives under the bark of the white oak, appears in June and July. The Chinch bug begins its terrible ravages in the wheat fields. The various species of Chrysopa or Lace-winged flies, appear during this month.

## The Insects of July.

During mid-summer the bees and wasps are very busy building their nests and rearing their young. The Humble bees, late in June and the first of this month, send out their first broods of

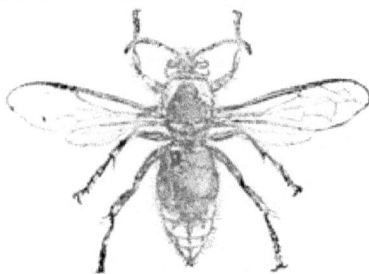

252. White-faced Wasp.

workers, and about the middle of the month the second lot of eggs are laid, which produce the smaller-sized females and males, while eggs laid late in the month and early in August, produce the larger-sized queens, which soon hatch. These hibernate. The habits of their peculiar parasite, Apa-thus, an insect which closely resembles the Humble bee, are still unknown.

The Leaf-cutter bee (Megachile) may be seen flying about with pieces of rose-leaf, with which she builds, for a period of twenty days, her cells, often thirty in number, using for this purpose, according to Mr. F. W. Putnam's estimate,* at least one thousand pieces! The bees referred to "worked so diligently that they ruined five or six rose-bushes, not leaving a single unblighted leaf uncut, and were then forced to take the leaves of a locust tree as a substitute."

The Paper-making wasps, of which Vespa maculata (Fig. 252),

*See "Proceedings of the Essex Institute," vol. iv, p. 105.

the "White-faced wasp," is our largest species, are now completing their nests, and feeding their young with flies. The Solitary wasp (Odynerus albophaleratus) fills its earthen cells with minute caterpillars, which it paralyzes with its poisonous sting. A group of mud-cells, each stored with food for the single larva within, we once found concealed in a deserted nest of the American Tent caterpillar. Numerous species of Wood wasps (Crabronidæ) are engaged in tunnelling the stems of the blackberry, the elder, and syringa, and enlarging and refitting old nail holes, and burrowing in rotten wood, storing their cells with flies, caterpillars, aphides and spiders, according to the habit of each species. Eumenes fraterna, which attaches its single, large, thin-walled cell of mud to the stems of plants, is, according to Dr. T. W. Harris, known to store it with Canker worms. Pelopæus, the Mud-dauber, is now building its earthen cells, plastering them on old rafters and stone walls.

The Saw flies (Tenthredo), etc., abound in our gardens this month. The Selandria vitis attacks the vine, while Selandria rosæ, the Rose slug, injures the rose. The disgusting Pear slug-worm (S. cerasi), often

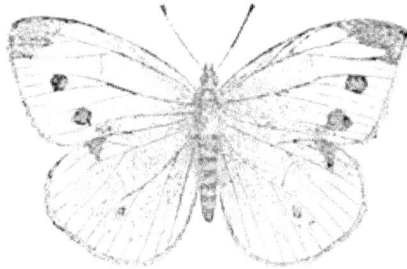

253. Imported Cabbage Butterfly.

live twenty to thirty on a leaf, eating the parenchyma, or softer tissues, leaving the blighted leaf. The leaves should be sprinkled with a mixture of whale-oil soap and water, in the proportion of two pounds of soap to fifteen gallons of water.

Among the butterflies, Melitæa Ismeria, in the south, and M. Harrisii, in the north, are sometimes seen. A second brood of Colias Philodice, the common sulphur-yellow butterfly, appears, and Pieris oleracea visits turnip-patches. It lays its eggs in June on the leaves, and the full-grown, dark-green, hairy larva may be found in August. The Pieris rapæ, or imported cabbage butterfly (Fig. 253, male) is now also abundant. Its green hairy larva is fearfully prevalent about Boston and New York. The last of the month a new brood of Grapta comma appears, and a second brood of the larva of Chrysophanus Americanus may be found on the sorrel.

The larvæ of Pyrraretia Isabella hatch out the first week in

July, and the snuff-colored moth enters our windows at night, in company with a host of night-flying moths.   These large

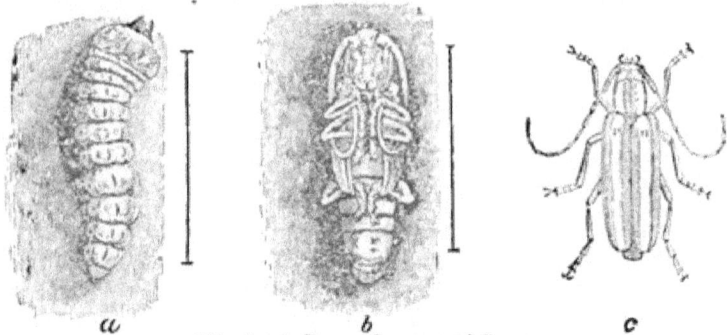

251.  Apple Borer, Larva and Pupa.

moths, many of which are injurious to crops, are commonly thought to feed on clothes and carpets.  The true carpet and clothes moths are minute species, which flutter noiselessly about our apartments.  Their narrow, feathery wings are edged with long silken fringes, and almost the slightest touch kills them.

255. Lady Bug and Pupa.

Among beetles, the various borers, such as the Saperda, or apple tree borer (Fig. 254) are now pairing. and fly in the hot sun about trees. Nearly each tree has its peculiar enemy, which drives its galleries into the trunk and branches of the tree.   Among the Tiger beetles, frequenting sandy places, the large Cicindela

256. Lace-winged Fly and Eggs.

generosa and the Cicindela hirticollis are most common.  The grotesque larvæ live in deep holes in sand-banks.

The nine-spotted Lady Bug, Coccinella novemnotata (Fig. 255. with pupa) is one of a large group of beetles. most beneficial from their habit of feeding on the plant lice. We figure another enemy of the

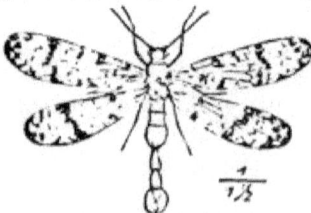

257. Forceps-tail.

Aphides, Chrysopa, and its eggs (Fig. 256). mounted each on a long silken stalk, thus placed above the reach of harm.

Among other beneficial insects belonging to the Neuroptera, is the immense family of Libellulidæ, or Dragon flies. The Forceps-tail, or Panorpa, P. rufescens (Fig. 257), is found in bushy fields and shrubbery. They prey on smaller insects, and the males are armed at the extremity of the body with an enormous forceps-like apparatus.

## The Insects of August.

During this month great multitudes of bugs (Hemiptera) are found in our fields and gardens; and to this group of insects the present chapter will be devoted. They are nearly all injurious to crops, as they live on the sap of plants, stinging them with their long suckers. Their continued attacks cause the leaves to wither and blight.

The grain Aphis, in certain years, desolates our wheat fields. We have seen the heads black with these terrible pests. They

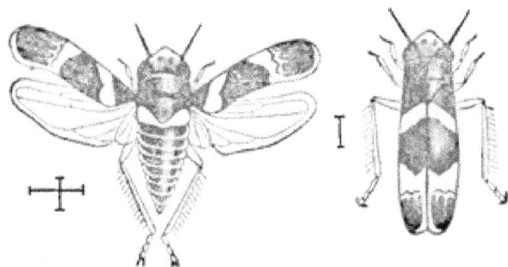

258. Leaf-hopper of the Vine.

pierce the grain, extract the sap, causing it to shrink and lose the greater part of its bulk. It is a most insidious and difficult foe to overcome.

The various leaf-hoppers, Tettigonia (Fig. 258) and Ceresa, abound on the leaves of plants, sadly blighting them; and the Tettigonias frequent damp, wet, swampy places. A very abundant species on grass produces what is called "frog's spittle." It can easily be traced through all its changes by frequently examining the mass of froth which surrounds it. Tettigonia vitis blights the leaf of the grape-vine. It is a tenth of an inch long, and is straw-yellow, striped with red. Tettigonia rosæ, a still smaller species, infests the rose, often to an alarming extent.

The Notonecta, or water boatman, is much like a Tettigonia, but its wings are transparent on the outer half, and its legs are

fringed with long hairs, being formed for swimming.  It rows
over the surface in pursuit of insects.  Notonecta undulata Say
(Fig. 259) is a common form in New England.

Another insect hunter is the singular Ranatra fusca (Fig. 260).
It is light brown in color, with a long respiratory tube which it
raises above the surface of the water when it wishes to breathe.
This species connects the Water-boatman with the Water-

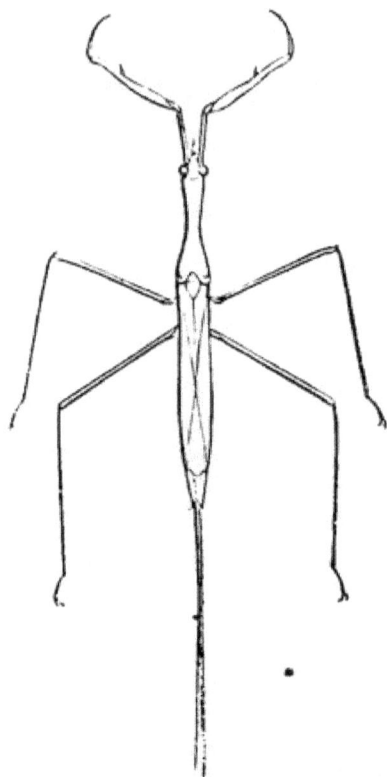

259. Notonecta.

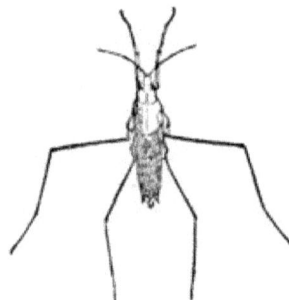

261. Water Skater.

260. Ranatra.

262. Pirates.

skaters, or Gerris, a familiar insect, of which Gerris paludum
(Fig. 261) is commonly seen running over the surface of streams
and pools.

Reduvius and its allies belong to a large family of very useful
insects, as they prey largely on caterpillars and noxious insects.
Such is Pirates picipes (Fig. 262), a common species.  It is an

ally of Reduvius personatus, a valued friend to man, as in Europe it destroys the bed-bug. Its specific name is derived from its habit while immature, of concealing itself in a case of dust, the better to approach its prey.

Another friend of the agriculturist is the Phymata erosa (Fig. 263). Mr. F. G. Sanborn states that "these insects have been taken in great numbers upon the linden trees in the city of Boston, and were seen in the act of devouring the Aphides, which have infested the shade trees of that city for several years past. They are described by a gentleman who watched their operations with great interest, as 'stealing up to a louse, coolly seizing and tucking it under the arm, then inserting the beak and sucking it dry.' They are supposed to feed also on other vegetable-eating insects as well as the plant louse."

263. Phymata.

Phytocoris lineolaris swarms in our gardens during this month. It is described and figured in "Harris's Treatise on Insects." Closely allied, though generally wingless, is that enemy of our peace, the bed-bug. It has a small, somewhat triangular head, orbicular thorax, and large, round, flattened abdomen. It is generally wingless, having only two small wing-pads instead. The eggs are oval, white; the young escape by pushing off a lid at one end of the shell. They are white, transparent, differing from the perfect insect in having a broad, triangular head, and short, thick antennæ. Indeed, this is the general form of lice (Pediculus vestimenti, and P. capitis), to which the larva of Cimex has the closest affinity. Some Cimices are parasites, infesting pigeons, swallows, etc., in this way also showing their near relation to lice. Besides the Reduvius, the cockroach is the natural enemy of the bed-bug, and destroys large numbers. Houses have been cleared of bugs after being thoroughly fumigated with brimstone.

During this month the ravages of grasshoppers are, in the West, very wide-spread. We have received from Major F. Hawn, of Leavenworth, Kansas, a most interesting account of the Red-legged locust (Caloptenus femur-rubrum). "They commence depositing their eggs in the latter part of August. They are fusiform, slightly gibbous, and of a buff-color. They are placed about three-fourths of an inch beneath the surface, in a compact mass around a vertical axis, pointing obliquely up and outwards, and are partially cemented together, the whole presenting a cylindrical structure, not unlike a small cartridge. They commence hatching in March, but it requires a range of temperature above 60° F. to bring them to maturity, and under such conditions they become fledged in thirty-three days, and in from three to five days after they enter upon their migratory flight.

"Their instincts are very strong. When food becomes scarce at one point, a portion of them migrate to new localities, and this movement takes place simultaneously over large areas. In

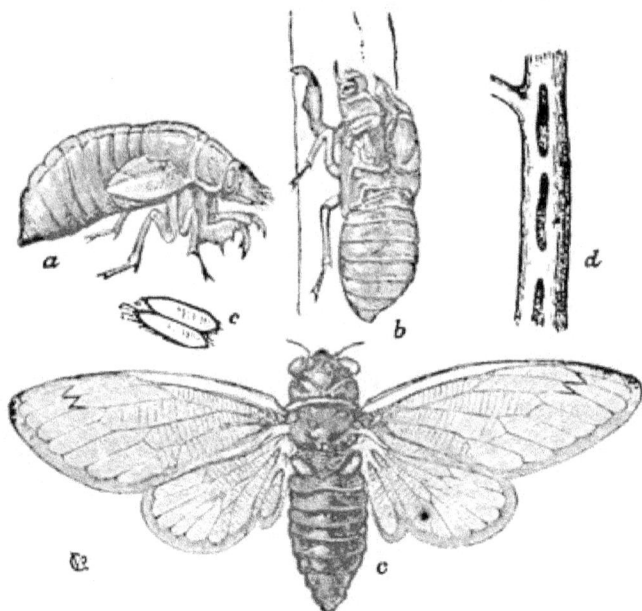

264. Seventeen Year Locust, Eggs and Pupa.

their progress they stop at no obstacle they can surmount. In these excursions they often meet with other trains from an opposite direction, when both join in one.

"The insects are voracious, but discriminating in their choice of food, yet I know of no plant they reject if pressed by hunger; not even the foliage of shrubs and trees, including pine and cedar."

During this month the Seventeen-year locust (Cicada septendecim of Linnæus, Fig. 264) has disappeared, and only a few Harvest flies, as the two other species we have are called, raise their shrill cry during the dog-days. But as certain years are marked by the appearance of vast swarms in the Middle States, we cannot do better than to give a brief summary of its history, which we condense in part from Dr. Harris' work.

The Seventeen-year locust ranges from South-eastern and Western Massachusetts to Louisiana. Of its distribution west of the Mississippi Valley, we have no accurate knowledge. In Southern Massachusetts, they appear in oak forests about the middle of June. After pairing, the female, by means of her powerful ovipositor, bores a hole obliquely to the pith, and lays therein from ten to twenty slender white eggs, which are arranged in pairs, somewhat like the grains on an ear of wheat, and implanted in the limb. She thus oviposits several times in a twig, and passes from one to another, until she has laid four or five hundred eggs. After this she soon dies. The eggs hatch in about two weeks, though some observers state that they do not hatch for from forty to over fifty days after being laid. The active grubs are provided with three pairs of legs. After leaving the egg they fall to the ground, burrow into it, and seek the roots of plants whose juices they suck by means of their long, beaks. They sometimes attack the roots of fruit trees, such as the pear and apple. They live nearly seventeen years in the larva state, and then in the spring change to the pupa, which chiefly differs from the larva by having rudimentary wings. The damage done by the larvæ and pupæ, then, consists in their sucking the sap from the roots of forest, and occasionally fruit trees.

Regarding its appearance, Mr. L. B. Case writes us (June 15) from Richmond, Indiana: "Just now we are having a tremendous quantity of locusts in our forests and adjoining fields, and people are greatly alarmed about them; some say they are Egyptian locusts, etc. This morning they made a noise, in the woods about half a mile east of us, very much like the continuous sound of frogs in the early spring, or just before a storm at evening. It lasted from early in the morning until evening." Mr. V. T. Chambers writes us that it is abounding in the vicinity of Covington, Kentucky, "in common with a large portion of the Western country." He points out some variations in color from those described by Dr. Fitch, from New York, and states that those occurring in Kentucky are smaller than those of which the measurements are given by Dr. Fitch, and states that "these differences indicate that the groups, appearing in different parts of the country at intervals of seventeen years, are of different varieties." A careful comparison of large numbers collected from different broods, in different localities, and different years, would alone give the facts to decide this interesting point. Mr. Riley has shown that in the Southern States a variety appears every thirteen years.

Regarding the question raised by Mr. Chambers, whether the sting of this insect is poisonous, and which he is inclined to believe to be in part true, we might say that naturalists generally believe it to be harmless. No hemiptera are known to be poisonous, that is, to have a poison-gland connected with the sting, like that of the bee, and careful dissections by the eminent French naturalist, Lacaze-Duthiers, of three European species of Cicada, have not revealed any poison apparatus at the base of the sting. Another proof that it does not pour poison into the wound made by the ovipositor is, that the twig thus pierced and wounded does not swell, as in the case of plants wounded by Gall flies, which, perhaps, secrete an irritating poison, giving rise to tumors of various shapes. Many insects sting without poisoning the wound; the bite of the mosquito, black fly, flea, the bed bug, and other hemipterous insects, are simply punctured wounds, the saliva introduced being slightly irritant, and to a perfectly healthy constitution they are not poisonous, though they may grievously afflict some persons, causing the adjacent parts to swell, and in some weak constitutions induce severe sickness. Regarding this point, Mr. Chambers writes: "I have heard—not through the papers—within a few days past of a child, within some twenty miles of this place, dying from the sting of a Cicada, but have not had an opportunity to inquire into the truth of the story, but the following you may rely on. A negro woman in the employment of A. V. Winston, Esq., at Burlington, Boone County, Ky., fifteen miles distant from here, went barefooted into his garden a few days since, and while there was stung or bitten in the foot by a Cicada. The foot immediately swelled to huge proportions, but by various applications the inflammation was allayed, and the woman recovered. Mr. Winston, who relates this, stands as high for intelligence and veracity as any one in this vicinity. I thought, on first hearing the story, that probably the sting was by some other insect, but Mr. Winston says that he saw the Cicada. But perhaps this proves that the sting is *not* fatal; that depends on the

subject. Some persons suffer terribly from the bite of a mosquito, while others scarcely feel them. The cuticle of a negro's
foot is nearly impenetrable, and perhaps the sting would have
been more dangerous in a more tender part." It is not improbable that the sting was made by a wasp (Stizus) which preys on
the Cicada. Dr. Le Baron and Mr Riley believe the wound to
be made by the beak, which is the more probable solution of the
problem.

A word more about the Seventeen-year Cicada. Professor
Orton writes us from Yellow Springs, Ohio, that this insect

265. Hop Vine Moth and Young. .

has done great damage to the apple, peach, and quince trees,
and is shortening the fruit crop very materially. By boring
into twigs bearing fruit, the branches break and the fruit goes
with them. "Many orchards have lost full two years' growth.
Though the plum and
cherry trees seemed ex-
empt, they attacked the
grape, blackberry, rasp-
berry, elm (white and
slippery), maple, white
ash, willow, catalpa,
honey-locust and wild
rose. We have traces

266. Humble Bee Parasite.

of the Cicada this year from Columbus, Ohio, to St. Louis.
Washington and Philadelphia have also had a visitation."

We figure the Hop-vine moth and the larva (Fig. 265)
which abound on hops the last of summer. Also, the Ilythia
colonella (Fig. 266, a, pupa), known in England to be a parasite of the Humble bee. We have frequently met with it here,
though not in Humble bees' nests. The larvæ feed directly upon
the young bees, according to Curtis (Farm Insects). The
Spindle-worm moth (Gortyna zeæ), whose caterpillar lives in
the stalks of Indian corn, and also in dahlias, flies this month.

The withering of the leaves when the corn is young, shows the presence of this pest. The beetles of various cylindrical Bark borers and Blight beetles (Tomicus and Scolytus) appear again

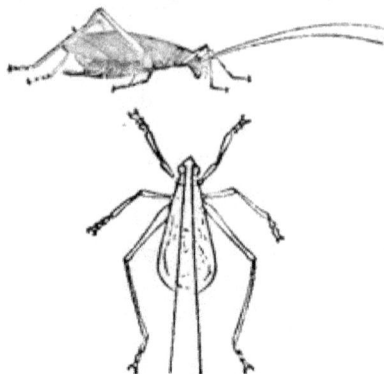

267. Tree Cricket.

this month. During this month the Tree cricket (Œcanthus niveus, Fig. 267) lays its eggs in the branches of peach trees. It will also eat tobacco leaves.

We figure (268) the moth of Ennomos subsignaria, the larva of which is so injurious to shade trees in New York City. It is a widely diffused species, occurring probably throughout the Northern States. We have taken the moth in

Northern Maine. We have received from Mr. W. V. Andrews the supposed larvæ of this moth. They are "loopers," that is, they walk with a looping gait, as if measuring off the ground they walk over, whence the name "Geometers," more usually applied to them. They are rather stout, brown, and roughened like a twig of the tree they inhabit, with an unusually large rust-red head,

268. Ennomos subsignaria.

and red prop-legs, while the tip of the body is also red. They are a little over an inch long.

### The Insects of September.

Few new insects make their first appearance for the season during this month. Most of the species which abound in the early part of the month are the August forms, which live until they are killed by the frosts late in the month. From this cause there is towards the end of the month a very sensible diminution of the number of insects.

The early frosts warn these delicate creatures of approaching cold. Hence the whole insect population is busied late in the month in looking out snug winter quarters, or providing for the continuance of the species. Warned by the cool and frosty nights, multitudes of caterpillars prepare to spin their dense silken cocoons, which guard them against frost and cold. Such are the "Spinners," as the Germans call them, the Silk moths, of which the American Silk worm is a fair example. The last of September it spins its dense cocoon, in which it hibernates in the chrysalis state.

The larvæ of those moths, such as the Sphinges, or Hawk moths, which spin no cocoon, descend deep into the earth, where they transform into chrysalids and lie in deep earthen cocoons.

The wild bees may now be found frequenting flowers in considerable numbers. Both sexes of the Humble bee, the Leaf-cutter bee, and other smaller genera abound during the warm days.

One's attention during an unusually warm and pleasant day in this month is attracted by clouds of insects filling the air, especially towards sunset, when the slanting rays of the sun shine through the winged hosts. On careful investigation these insects will prove to be nearly all ants, and, perhaps, to belong to a single species. Looking about on the ground, an unusual activity will be noticed in the ant-hills. This is the swarming of the ants. The autumnal brood of females has appeared, and this is their marriage day.

The history of a *formicarium*, or ant's nest, is as follows: The workers, only, hibernate, and are found early in the spring, taking care of the eggs and larvæ produced by the autumnal brood of females. In the course of the summer these eggs and larvæ arrive at maturity, and swarm on a hot sultry day, usually early in September. The females, after their marriage flight, for the small diminutive males seek their company at this time, descend and enter the ground to lay their eggs for new colonies, or, as Westwood states, they are often seized by the workers and retained in the old colonies. Having no more inclination to fly, they pluck off their wings and may be seen running about wingless.

Dr. C. C. Abbot gives us the following account of the swarming of a species in New Jersey: "On the afternoon of Oct. 6th, at about 4 P. M., we were attracted to a part of the large yard surrounding our home, by a multitude of large sized insects that filled the air, and appeared to be of some unusual form of insect life, judging of them from a distance. On closer inspection these creatures proved to be a brood of red ants (Formica) that had just emerged from their underground home and were now for the first time using their delicate wings. The sky, at the time, was wholly overcast; the wind strong, southeast; thermometer 66° Fahr. Taking a favorable position near the mass, as they slowly crawled from the ground, up the blades of grass and stems of clover and small weeds, we noted, first, that·they seemed dazed, without any method in their movements, save an ill-defined impression that they must go somewhere. Again, they were pushed forward, usually by those coming after them, which seemed to add to their confusion. As a brood or colony of insects, their every movement indicated that they were wholly ill at ease.

"Once at the end of a blade of grass, they seemed even more puzzled as to what to do. If not followed by a fellow ant, as was usually the case, they would invariably fall down again to the earth, and sometimes repeat this movement until a new comer joined in the ascent, when the *uncertain* individual would be forced to use his wings. This flight would be inaugurated by a very rapid buzzing of the wings, as though to dry them, or prove their owner's power over them, but which it is difficult to say. After a short rest, the violent movement of the wings would recommence, and finally losing fear, as it were, the ant would let go his hold upon the blade of grass and rise slowly upwards. It could, in fact, scarcely be called flight. The steady vibration of the wings simply bore them upwards, ten, twenty or thirty feet, until they were caught by a breeze, or by the steadier wind that was moving at an elevation equal to the height of the surrounding pine and spruce trees. So far as we were able to discover, their wings were of the same use to them, in transporting them from their former home, that the 'wings' of many seeds are, in scattering them; both are wholly at the mercy of the winds.

"Mr. Bates, in describing the habits of the Saüba ants (Œcodoma cephalotes) says,* 'The successful *début* of the winged males and females depends likewise on the workers. It is amusing to see the activity and excitement which reign in an ant's nest when the exodus of the winged individuals is taking place. The workers clear the roads of exit, and show the most lively interest in their departure, although it is highly improbable that any of them will return to the same colony. The swarming or éxodus of the winged males and females of the Saüba ant takes place in January and February, that is, at the commencement of the rainy season. They come out in the evening in vast numbers, causing quite a commotion in the streets and lanes.' We have quoted this passage from Mr. Bates' fascinating book, because of the great similarity and dissimilarity in the movements of the two species at this period of their existence. Remembering, at the time the above remarks concerning the South American species, we looked carefully for the workers, in this instance, and failed to discover above half a dozen wingless ants above ground, and these were plodding about, very indifferent, as it appeared to us, to the fate or welfare of their winged brothers. And on digging down a few inches, we could find but comparatively few individuals in the nest, and could detect no movements on their parts that referred to the exodus of winged individuals, then going on.

"On the other hand, the time of day agrees with the remarks of Mr. Bates. When we first noticed them, about 4 P. M., they had probably just commenced their flight. It continued until nearly 7 P. M., or a considerable time after sundown. The next morning, there was not an individual, winged or wingless, to be seen above ground; the nest itself was comparatively empty; and what few occupants there were seemed to be in a semi-torpid condition. Were they simply resting after the fatigue and excitement of yesterday?

"It was not possible for us to calculate what proportion of these winged ants were carried by the wind too far to return to their old home; but certainly a large proportion were caught by the surrounding trees; and we found, on search, some of these crawling down the trunks of the trees, with their wings in a damaged condition. How near the trees must be for them to reach their old home, we should like to learn; and what tells them, 'which road to take?' Dr. Duncan states,* 'It was formerly supposed that the females which alighted at a great distance from their old nests returned again, but Huber, having

---

*Transformations of Insects, p. 205.

great doubts upon this subject, found that some of them, after having left the males, fell on to the ground in out-of-the-way places, whence they could not possibly return to the original nest!' We unfortunately did not note the sex of those individuals that we intercepted in their return (?) trip; but we can not help expressing our belief that, at least in this case, there was scarcely an appreciable amount of 'returning' on the part of those whose exodus we have just described; although so many were caught by the nearer trees and shrubbery. Is it probable that these insects could find their way to a small underground nest, where there was no 'travel' in the vicinity, other than the steady departure of individuals, who, like themselves, were terribly bothered with the wings they were carrying about with them?" (*American Naturalist.*)

We have noticed that those females that do not return to the old nest found new ones. In Maine and Massachusetts we have for several successive years noticed the swarming of certain species of ants during an unusually warm and sultry day early in September.

The autumnal brood of Plant lice now occur in great numbers on various plants. The last brood, however, does not consist exclusively of males and females, for of some of the wingless individuals previously supposed to be perfect insects of both sexes, Dr. W. I. Burnett found that many were in reality of the ordinary gemmiparous form, such as those composing the early summer broods.

The White Pine Plant lice (Lachnus strobi) may be seen laying their long string of black oval eggs on the needles of the pine.* They are accompanied by hosts of two-winged flies, Ichneumons, and in the night by many moths which feed on the Aphis-honey they secrete, and which drops upon the leaves beneath.

# INDEX.

the demands of both, and through the aid of an unusually large number of excellent woodcuts, it affords the greatest possible aid in the definition and means of identifying the specimens in hand. That a work which should be useful to farmers and gardeners should treat especially of the useful and injurious insects is evident. This part of entomology has within the last ten years made great progress. Packard has here also understood how to preserve a right proportion in the choice and handling of his material. * * We may consider Packard's work as among the best guides to the study of insects which at present exist. We close our notice with the hope that the book may find its way in Europe as rapidly as it seems to have in America. — *Translation from the Entomological Journal of Stettin, Prussia.*

It fills a gap in American entomological literature, is well written and amply illustrated, and we hope will meet with the success it deserves. The work will be found an invaluable guide to the study of insects.—*The American Entomologist.*

A thoroughly good, reliable, well illustrated manual of structural and systematic entomology, prepared by one who is a master in the science.—*The Canadian Entomologist.*

Dr. Packard's GUIDE is a work which we trust will find its way to all who take any interest, scientific or practical, or both, in the world of insects. It is designed to assist the student in learning the structure, transformations and development of insects, to direct him in their collection and preservation, and to help him in the identification of his captures. It is also intended to afford a useful account of those classes of insects that are especially injurious to vegetation, and at the same time draw attention to those that are beneficial.—*Canada Farmer.*

The study of Entomology is one that can be almost as easily pursued in the school-room as Botany, and we should be glad to see the experiment tried. Dr. Packard's work is well fitted, by its clear, simple style, for use as a text-book.—*The Michigan Teacher.*

The insects injurious and beneficial to vegetation are especially noticed. In typography and illustrations it leaves nothing to be desired, while the author's style is very perspicuous. We hope the work will be used as intended in colleges and farm-schools, and by agriculturists.—*The Nation.*

Altogether, we are immensely pleased with this work. It is assuredly all in all the fullest, most modern, and most clearly-written treatise on insects we have ever seen, and we heartily commend it to our readers' notice, feeling certain their judgment of its merits will not be less favorable than our own.—*Popular Science Review,* London.

The first portion of this book, occupying nearly two parts, is devoted to general entomology, and furnishes an admirable, though necessarily brief, account of their organization, of their reproduction and development in the egg and of their metamorphoses. The most recent memoirs connected with these subjects, have been made use of by the author, and this part of his work is certainly the best manual of entomology which the English reader can at present obtain.—*Nature,* London.

As a practical treatise on American Entomology, with reference especially to the insects injurious or beneficial to crops, it stands almost alone, and reflects the highest credit upon American scholarship, patience and scientific skill.—*New York Tribune.*

The "American Naturalist" is an elegantly printed and ably conducted magazine, which we have already had occasion to notice as from the press of the Peabody Academy, of Salem. One of its editors is the author of this GUIDE TO THE STUDY OF INSECTS, and the elegant typography and illustrations, so characteristic of that magazine, and the zeal and scholarship of Mr. Packard, makes this by far the best work in the language on this subject. The study of insects is becoming the most popular branch of natural history.—*New York Independent.*

It is a treatise which any intelligent farmer may study with pleasure and profit, and one which will at the same time greatly edify even the learned naturalist.—*New York Observer.*

# THE NATURALISTS' AGENCY.

THE many letters received by the Editors of the *Naturalist* requesting information about obtaining books and papers on TOPICS OF NATURAL HISTORY induced them to establish an Agency at the office of the *American Naturalist*, for the purpose of SUPPLYING NATURALISTS WITH BOOKS AND PAMPHLETS, and aiding Authors and Institutions in selling their various publications on the following plan:—

Parties to send to the Agency several copies (not exceeding twenty, unless specially requested) of each book or pamphlet they have for sale, stating the *retail price*.

An account will be opened with each party sending, and credit given at the retail price for all books received.

Parties will be charged on their accounts for any transportation expenses paid on receiving the packages, which should be sent by mail prepaid, when not too bulky.

A classified list of the books will be given from time to time in the *Naturalist*, with the prices of each annexed, and the amount to be remitted in addition for postage.

Yearly accounts of each party will be made up, and a statement forwarded regarding the number of copies on hand, and the sales that have been made, for which the cash will be remitted after deducting any express charges incurred on the receipt of the books, and the commission to defray the expenses of the advertisement, etc.

Parties having works on Natural History which they wish to dispose of can send them to the Agency on the same terms.

The Agency list also contains notices of the various Scientific Periodicals, and the regular publications of Scientific Societies and Institutions that have made us their agents.

It will be part of the business of the Agency to procure, if possible, any works on Natural History, other than those on its list, that may be ordered.

*To Institutions and Societies publishing Proceedings, Transactions, Memoirs, etc.*, the Agency is proving of great benefit, and AUTHORS HAVING EXTRA COPIES OF THEIR OWN PAPERS FOR SALE are finding it to their advantage in thus making their papers more widely known.

It is specially requested that all packages for the Agency be addressed as follows.

## The Naturalists' Agency,

[Mark legibly from whom the package comes.]

### SALEM, MASS.

About one hundred writers on Natural History now supply the public with their works through the *Naturalists' Agency.*

*⁎* Publications of works on Natural History undertaken for authors.

The Agency also keeps on hand a supply of Insect Pins, Cork for Insect Boxes, dissecting Knives, Forceps, and other instruments in general use by Naturalists. Also,— Bird's Skins and Eggs, Shells, Insects, etc., etc., that have been consigned for sale. Send for Catalogue.

#### W. S. WEST, Agent.

SALEM, MASS.

## THE SECOND EDITION (REVISED AND CORRECTED) OF

HORACE MANN'S CATALOGUE OF PLANTS. (The Phænogamous and Vascular Cryptogamous Plants of the United States East of the Mississippi.)

PRICE 35 CENTS.    THREE FOR $1.00.

www.ingramcontent.com/pod-product-compliance
Lightning Source LLC
Chambersburg PA
CBHW021527210326
41599CB00012B/1413